Principles of the
Human Genome and
Pharmacogenomics

Principles of the Human Genome and Pharmacogenomics

Daniel A. Brazeau, PhD
Gayle A. Brazeau, PhD

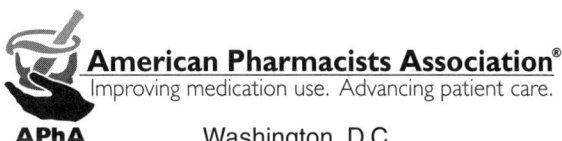

American Pharmacists Association®
Improving medication use. Advancing patient care.

APhA   Washington, D.C.

Managing Editor: Vicki Meade, Meade Communications
Acquiring Editor: Sandra Cannon
Copy Editor: Deborah J. Shuman
Proofreader: Betty Bruner
Indexer: Jennifer Burton, Columbia Indexing Group
Cover Designer: Mariam Safi, APhA Creative Services
Layout and Graphics: Michele A. Danoff, Graphics by Design

Published by the American Pharmacists Association
2215 Constitution Avenue, N.W.
Washington, DC 20037-2985
www.pharmacist.com
www.pharmacylibrary.com

To comment on this book via e-mail, send your message to the publisher at
aphabooks@aphanet.org

Library of Congress Cataloging-in-Publication Data

Brazeau, Daniel A.
  Principles of the human genome and pharmacogenomics / Daniel A Brazeau
and Gayle A. Brazeau.
      p. ; cm.
  Includes bibliographical references and index.
  ISBN 978-1-58212-124-6
  1. Pharmacogenetics. 2. Pharmacogenomics. 3. Human genetics. I.
Brazeau, Gayle A. II. American Pharmacists Association. III. Title.
  [DNLM: 1. Pharmacogenetics--methods. 2. Genome, Human--drug effects. QV 38]
  RM301.3.G45B73 2011
  615'.7--dc22
                            2010049353

How to Order This Book
Online: www.pharmacist.com/shop_apha
By phone: 800-878-0729 (from the United States and Canada)
VISA®, MasterCard®, and American Express® cards accepted

Contents

Preface

The use of pharmacogenetics and pharmacogenomics in clinical practice and the pharmaceutical sciences encompasses much more than a list of drugs and the known genes that are their therapeutic targets or that play a role in their metabolism and distribution. The application of genetic and genomic principles to our understanding of human health is revolutionizing health care and practice standards for health care providers in all professional settings, and it will have a major impact on future generations of clinical, biomedical, and pharmaceutical scientists.

Drugs and diagnostics in the post–human genome years will be directed to specific molecular targets, probably more so than to specific genotypes. It will become increasingly difficult to practice contemporary pharmacy and to provide satisfactory pharmacist care in the future without a fundamental knowledge of genetics and genomics, because most important interactions with patients and other health care professionals will require pharmacists to have a working comprehension of the human genome and pharmacogenomics.

The purpose of this book is to introduce readers—both students and practitioners—to important principles of human genetics and genomics that they can apply and integrate in the management of medication therapy for their patients. It is not our intent to provide the reader with a synopsis of known pharmacogenetic genes and their drugs. Rather, we have attempted to give professional and student pharmacists a concise source of the critical science underlying the basics of pharmacogenetics and pharmacogenomics.

A strong scientific foundation is essential for pharmacists to take a leadership role in the interprofessional care of patients, enabling them to serve as key members of the health care team. Understanding genomic science will be necessary for those who wish to stay abreast of contributions that the field of genetics will make to their discipline as the field advances in years to come. The challenges will be enormous, but health care professionals who have a strong scientific foundation have always been the ones to meet these challenges and to realize the opportunities in a changing world. Pharmacists in all settings are well situated to assume leadership roles in ensuring that advances made in pharmacogenomics and pharmacogenetics serve to optimize the care provided to their patients.

Each chapter begins with a list of learning outcomes summarizing important concepts the learner will have mastered by the chapter's end through thoughtful and critical reading. Important terms appear in boldface when additional information about them is found in the nearby text boxes, and these terms also appear in a glossary at the end of the book. Questions follow at the end of each chapter to further challenge the learner. In addition, key references appear at the end of each chapter, but the learner is cautioned to keep current with the literature in the biomedical, pharmacogenomic, and pharmaceutical sciences and in clinical practice. As pharmacogenomics and pharmacogenetics advance further, they will have an increased impact on professional practice for pharmacists and for all health care professionals.

Daniel A. Brazeau
Gayle A. Brazeau
November 2010

Acknowledgment

We gratefully acknowledge the patience and understanding of Sandy Cannon, who has waited far too long for us to finish this book.

CHAPTER 1

Introduction: Pharmacogenomics and Pharmacogenetics—A Historical Look

LEARNING OUTCOMES:

At the end of the chapter, you should be able to:

1. Identify and discuss significant historical findings as they relate to pharmacogenomic and pharmacogenetic issues in current contemporary pharmacy practice.

2. Identify and discuss the specific mechanisms by which genetic differences might account for differences in therapeutic effectiveness or toxicity as applied to patient care.

3. Discuss the two aspects of "personalized medicine."

For nearly a generation, it has been understood that some of the differences in how individuals respond to drugs are inherited and therefore, at least in part, genetic. Recent technological advances in genetics now allow pharmacists and other health care professionals to explain and anticipate some of this genetic variation in drug response. The rapidly emerging science of **pharmacogenomics** has the ultimate goal of identifying the many underlying genetic factors that play a role in the efficacy or toxicity of all drugs.

Pharmacogenomics is one of the most rapidly growing fields of biomedical science and is becoming integral to all aspects of drug discovery, design, and development. The science of pharmacogenomics represents the union of three fields of genetics—molecular, population, and quantitative genetics. Although it is not yet clear whether this pharmacogenomic revolution will have widespread clinical relevance, there is no doubt that in the future, health care professionals in general, and pharmacists in particular, will require significant understanding of genetics and genomics.

1.1 Genetics and Pharmacogenetics: A Brief History

Individuals have always differed in how they respond to drugs. The ways in which patients respond to a particular drug is too often unpredictable; responses range from little or no therapeutic benefit to harmful adverse drug reactions.[1,2] This lack of predictability in patient response results in substantial costs to contemporary health care systems. Many factors, including age, gender, drug interactions, and concomitant diseases and therapies, have long been known to affect treatment efficacy or toxicity (Figure 1-1). In addition, drug response may be enhanced or altered by drugs that a patient may be taking concurrently.[3]

Recently, the advent of molecular genetics and the dramatic development of genomic technologies have made it possible to consider the effect of a patient's underlying genetic makeup on drug response. This is true despite the massive complexity of the human **genome**, which has more than three billion gene-encoding "letters." However, neither the term **pharmacogenetics** nor the field it represents are recent phenomena (Figure 1-2). The term itself was coined in 1959 by Vogel, a German geneticist, 94 years after an Augustinian priest, Gregor Johann

Pharmacogenomics:
The study of the genome-wide role of human variation in drug response. Pharmacogenomics is a broad term that includes pharmacogenetic effects. Pharmacogenomics also includes the application of genomic technologies in drug discovery, disposition, and function.

Genome:
The genome of an organism encompasses all the genetic material in the cell. In humans, this includes the 3 billion base pairs contained in the chromosomes in the nucleus and the approximately 16,000 base pairs of the mitochondrion.

Pharmacogenetics:
The study of the role of genetic variation in determining individual drug response. Generally, pharmacogenetics has been limited to the effects of one or a few genes.

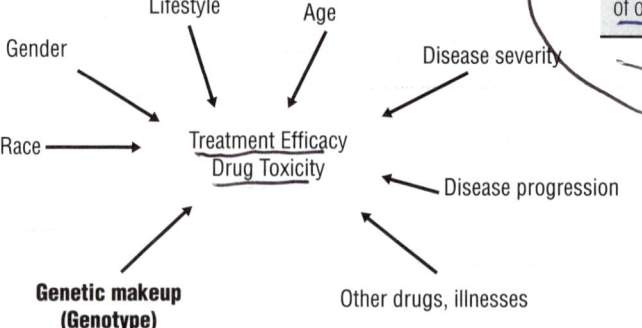

Figure 1-1. Many factors, working alone or in concert, determine how any given individual will respond to a drug. With the advent of genomic technologies and knowledge, it is becoming possible to assess the role an individual's genetic makeup or genotype plays in this response.

Pharmacogenomics		Genetics
	1859	Darwin publishes *On the Origin of Species*
	1865	Gregor Mendel's work published
	1869	Miescher isolates DNA
	1900	Rediscovery of Mendel's work by Carl Correns, Erich von Tschermak-Seysenegg, and Hugo DeVries
Chemical Individuality—"individuals do not conform to rigid standards of metabolism", Garrod.	1902	Bateson coins the term genetics and discovers linkage; also coins epistasis
	1910	Morgan's description of sex-linkage in *Drosophila melanogaster*; chromosomal basis of inheritance
Garrod's *Inborn Factors in Diseases* —first link between genetics and pharmacology—postulated a mutation in gene for enzyme responsible for metabolism	1931	
Snyder—PTC and "taste blindness" in 800 families	1932	
	1941	Beadle & Tatum—one gene, one enzyme (biochemical genetics and medicine)
	1944	Avery, MacCloed, and McCarty—DNA may be the stuff
Alving et al.—acute hemolytic response to primaquine in African American soldiers	1952	Hershey and Chase—DNA is the stuff
	1953	Watson and Crick's molecular model for DNA
Kalow and genetics of serum cholinesterase, first to show heritable variant enzyme and drug sensitivity Motulsky—*Drug reactions, enzyme, and biochemical genetics*	1957	
Vogel proposes the term *Pharmacogenetics*	1959	
Kalow publishes first monograph in pharmacogenetics	1962	
	1966	Genetic code described
	1970	Isolation of first restriction enzyme
Smith—Debrisoquine and CYP2D6	1977	DNA sequencing
	1985	Polymerase chain reaction
First cloning of "pharmacogenetic gene", CYP2D6, Gonzalez et al.	1988	Human Genome Project
	2000	First draft Human Genome

Figure 1-2. Genetic timeline showing the correspondence among discoveries in the pharmaceutical sciences and genetics. With our better understanding of genetics and the molecular biology of drug action, these two seemingly disparate fields are beginning to merge and hasten the advance of both.

Mendel, first described the laws that govern the inheritance of simple traits in pea plants. *Pharmacogenetics* has traditionally been defined as the study of the influence of a single gene on drug response. *Pharmacogenomics*, a more recent term, is often used interchangeably with *pharmacogenetics*, though the former is a broader term that includes not only the effects of a single gene but also the genome-wide influence on drug response, efficacy, and toxicity. The field of pharmacogenomics also includes the application of genomic technologies to identify networks of genes that affect drug efficacy and toxicity, ascertain new therapeutic drug targets, and optimize current pharmacotherapeutic treatments. It is this latter area of pharmacogenomics that is having an enormous, though largely unreported, impact on the pharmaceutical and biomedical sciences.

Perhaps the first link between genetics and pharmacology was made by Sir Archibald Garrod, who postulated in his book *Inborn Errors of Metabolism*[4] that a **mutation** in a gene coding for an enzyme may be responsible for human differences in the metabolism of drugs and environmental chemicals. The first large-scale study documenting human variation in response to a chemical was conducted by L. H. Snyder.[5] Snyder investigated over 750 families and showed that "taste blindness," the inability of some individuals to taste the chemical phenylthiocarbamide, was inherited as an **autosomal-recessive** trait.

One of the first documented "pharmacogenetic stories" was isoniazid, first synthesized in 1912, which became the first line of treatment for tuberculosis in the 1950s. Isoniazid is metabolized in the liver via **acetylation**, and elimination for the metabolite is primarily renal. As isoniazid increased in clinical usage, it was quickly noted that some patients reported peripheral neuropathies, specifically numbness in the arms or legs, often accompanied by pain. These complications were attributed to the interaction of the drug with pyridoxine, or vitamin B_6—specifically, the depletion of vitamin B_6. By 1954, the complications were found to be specifically associated with patients exhibiting deficiencies of a specific enzyme, N-acetyltransferase.[6] Patients with genetic deficiencies of N-acetyltransferase-2 exhibited a low ability to degrade isoniazid to acetylisoniazid and were termed "slow acetylators." Ultimately it was found that approximately 50% of African Americans and Caucasians are slow acetylators, whereas rapid acetylators are more common

Mutation:
A change in the DNA sequence of the genome. Mutations occurring in the germ line are potentially heritable. Changes in DNA sequence are of two basic types: single-nucleotide polymorphisms (SNPs) or insertion/deletions (indels) that can be from one to millions of nucleotides in size.

Autosomal:
Genes or loci that reside on any chromosome other than the sex chromosomes (i.e., the X and Y chromosomes).

Recessive:
A property of one of two alleles. An allele is said to be recessive when its phenotype is masked or unseen when in combination with another allele. The other allele is said to be dominant. Recessive alleles need not be rare in a population or deleterious to the individual.

among Asians. Fast acetylators have been identified with drug half-lives that are two to four times shorter than those seen in slow acetylators, and these differences have had clinical consequences for a number of important drugs other than isoniazid, including procainamide, hydralazine, phenelzine, and salicylazosulfapyridine.[7]

An interesting case of interactions among genes, drugs, and ethnic origins was noted during World War II. The commonly used antimalarial drug primaquine was found to cause hemolytic disease in an unusually high number of African American soldiers. After the war, work done in the Alving laboratory at the University of Chicago showed that a poor response occurred in patients with glucose 6-phosphate dehydrogenase deficiency.[8,9] The gene for glucose 6-phosphate dehydrogenase is found on the X chromosome and is one of the most **polymorphic** in humans. The deficiency was found to be more common among Americans of African, Mediterranean, and Asian descent and presumably reached higher frequencies in these populations because it provided some resistance to malaria. The frequency of these low-activity alleles of glucose 6-phosphate dehydrogenase is highest among populations where malaria is prevalent. In this case, the gene placing patients at risk is not part of the drug's metabolizing pathway, nor is it the immediate target of the drug. This example provides a hint of the complexity of the many gene–gene interactions that are now familiar to scientists. Individuals with this genetic defect are also prone to hemolytic events due to other causes, such as infections and ingestion of fava beans (favism).

Another early example of genetic differences in drug biotransformation was elucidated by Kalow and colleagues, who in 1957 demonstrated that prolonged apnea in response to the muscle relaxant succinylcholine was due to inherited structural differences in the enzyme pseudocholinesterase.[10] This work was the first to demonstrate a link between heritable differences in an enzyme structure and drug response in patients. Recently, Lockridge has shown that this enzyme variant is due to a substitution of the nucleotide at position 209, which results in a change in amino acids from aspartic acid to glycine.[11] Approximately one in 3,500 Caucasians are **homozygous** for atypical forms of this gene.[12]

Acetylation:
A reaction that introduces an acetyl functional group into a chemical compound. Most proteins are modified by acetylation.

Polymorphic:
A gene or locus is polymorphic if there are differences among individuals in its DNA sequence or length. Generally, the specific difference must have a frequency of 5% in the population to be considered polymorphic.

Homozygous:
A locus or individual is said to be homozygous if the two alleles present are identical. Heterozygous individuals carry different alleles at the locus of interest.

By the late 1950s, enough "pharmacogenetic cases" existed that the American Medical Association invited the geneticist Arno Motulsky to summarize the known findings in a paper titled "Drug Reactions, Enzymes, and Biochemical Genetics."[13] It was two years later that Vogel coined the term *pharmacogenetics*.[14] The field of pharmacogenetics had begun, and by 1962, Kalow had published the first book in the field.[15]

Perhaps the most studied genetic polymorphism in a drug-metabolizing enzyme, cytochrome P450 2D6 (CYP2D6) results in an abnormal and extended drop in blood pressure in response to the no-longer-used antihypertensive debrisoquine. Work has shown that subjects can be readily grouped into two classes, "poor metabolizers" and "extensive metabolizers."[16] Poor metabolizers are deficient or lacking in this enzyme. They have also been found to have lower urinary concentrations of metabolite and higher plasma concentrations of parent drug than do normal individuals or extensive metabolizers. Another drug, the anti-arrhythmic spartein, is also metabolized by CYP2D6 and produces a similar response.[17]

This work has become significant for two reasons (see Kalow 2004 for an excellent historical review[18]). First, this polymorphism is fairly common, having allele frequencies of 5% to 10% among Caucasians in Europe and North America, and thus is likely to be clinically important. Second, CYP2D6 is known to metabolize many clinically important drugs, including β-adrenergic–blocking agents, antidepressants, and anti-arrhythmics. CYP2D6 was the first pharmacogenetically relevant gene to be cloned and sequenced and to have its "poor metabolizer" alleles characterized.[19] Recently, many more genetic variants have been identified and shown to have effects ranging from low or no enzyme activity to individuals with multiple copies of the gene. Interestingly, the frequency of the multicopy CYP2D6 gene was 29% in one Ethiopian study,[20] a likely indication of the geographic origin of this mutation.

With the near completion of the sequencing of the human genome in 2000, the broader impact of the field of pharmacogenomics has grown at an increasing rate. The list of pharmacologically relevant genes has greatly increased to include not only those encoding traditional drug-metabolizing enzymes but also genes coding for drug transporters, as well as the many genes coding for the targets of drug action. Pharmacogenetic data are now routine components of new drug investigations and applications. Pharmacogenetic information is included in required drug labeling for new drugs as appropriate based on required clinical testing. The U.S. Food and Drug Administration has recently approved genetic tests for a number of medications. The role of genetics in pharmacy and pharmacy practice has a long history, and there is little doubt that genomics has and will continue to have a significant impact on the science and practice of pharmacy.

1.2 The Role of Pharmacogenomics in Pharmacy and Pharmacy Practice

Genomics is changing the nature of medicine and health care. The ability of genomic technologies to generate data has far outpaced researchers' abilities to assimilate the information. In as little as 20 years, sequencing rates have increased from efforts requiring two days to garner 500 base pairs of DNA sequence to automated systems generating millions of base pairs of sequence in a single day. The once unimaginable idea that individuals could have their entire genome sequenced at a cost similar to that of routine medical tests is now anticipated in the very near future. It's been estimated that the cost of determining an individual's genome sequence is decreasing by a factor of 2 with each year; thus the cost of determining the entire genome sequence for an individual at birth will be feasible, though likely unnecessary. Similarly, many thousands of human genetic polymorphisms for genes encoding drug-metabolizing enzymes can now be assayed on a single DNA "chip." These same genomic technologies enable the assessment of gene expression for thousands of genes from as little as a few cells, allowing for the very precise measurement of gene expression for specific tissues within an organ or tumor. These technologies are revolutionizing research in all the life sciences.

1.2.1 Personalized Medicine: Two Perspectives

It is now becoming possible to examine and ultimately comprehend the nature of disease and drug action at the molecular level. This information will certainly be the foundation of humankind's ultimate understanding of the nature of human health and disease. New molecular diagnostic tools will allow health care professionals to characterize human disease into ever finer distinctions and categorizations. Personalized medicine will come to mean not just the right drug for the right individual, but the right drug for the specific disease type afflicting the specific individual. This "individualization" of disease will allow for many more specific and successful therapeutic interventions than are now possible.

The speed at which technologies have become available and the ease with which they are conducted make the challenge not so much determining a patient's genetic makeup or specific disease state as deciphering the massive amount of data presented. The field of bioinformatics has grown in parallel to genomics as a means to address the computational complexity associated with increasing genetic information. Similarly, the greatest challenge to the translation of genomics and pharmacogenomics from "bench to bedside" is the education of pharmacists and other health care professionals. This educational deficit has been noted both within the curricula of schools of pharmacy[21] and among practicing pharmacists.[22] The need is particularly acute because the science of pharmacogenomics is really a combination of three information-rich areas of genetics (Figure 1-3):

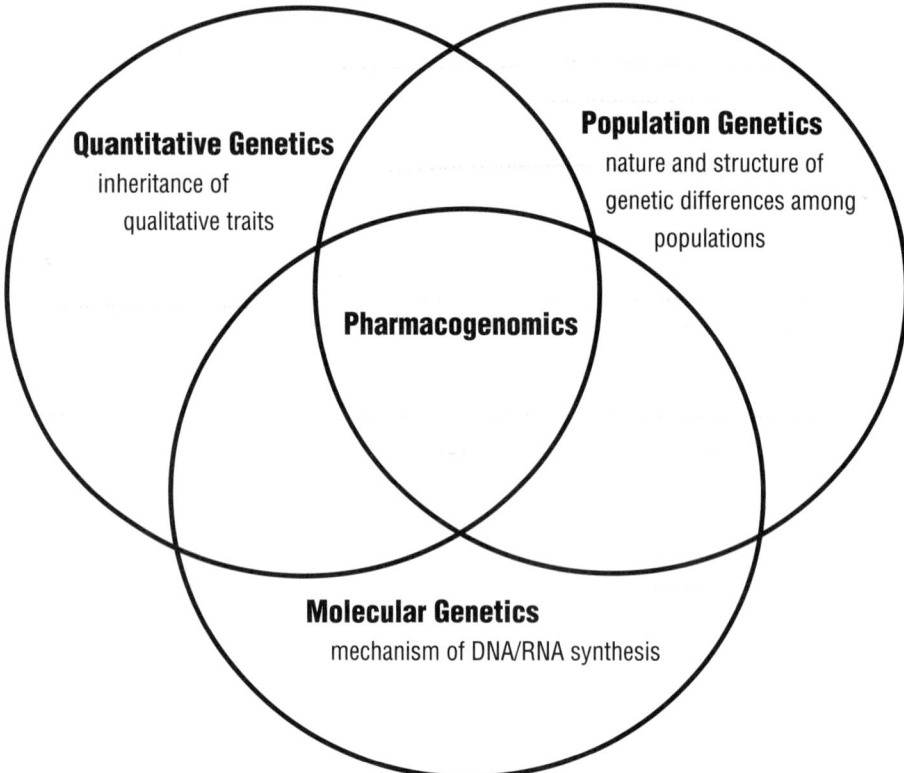

Figure 1-3. The field of pharmacogenomics/pharmacogenetics represents the intersection of three distinct disciplines in genetics. Ultimately, any "genomic" understanding of how a patient responds to a drug will be a function of the molecular mechanisms that underlie the cellular response (molecular genetics), the interaction between the patient's many genes and a multitude of environmental factors (quantitative genetics), and the variation in genetic background among human populations (population genetics).

- Molecular genetics—the study of the mechanisms of DNA and RNA synthesis, including gene regulation.
- Population genetics—the study of the nature, structure, and maintenance of genetic variation among populations.
- Quantitative genetics—the study of the inheritance of continuous or qualitative traits.

The future practice of pharmacy and medicine will require of its practitioners a basic understanding of genetic and genomic principles in order to realize the advancements that genomics offers.

QUESTIONS

1. What is a genome of an organism, and how does it relate to contemporary patient care?

2. Discuss the difference between the terms *pharmacogenetics* and *pharmacogenomics* as it relates to contemporary pharmacy practice and health care.

3. Identify specific types of polymorphisms that could affect the efficacy or toxicity of a specific drug.

4. Identify the significant findings made by the following individuals as they relate to differences in patients' response to drugs or disease:

    a. Sir Archibald Garrod
    b. L. H. Snyder
    c. A. S. Alving
    d. A. Motulsky
    e. W. Kalow
    f. R. L. Smith
    g. F. Vogel

5. Define "personalized medicine" and its significance in patient care.

6. Pharmacogenomics is the integration of what three genetic areas?

7. What genetic polymorphism is fairly common with respect to drug biotransformation, and what is the clinical significance in pharmacy practice?

References

1. Lazarou J, Pomeranz BH, Corey PN. Incidence of adverse drug reactions in hospitalized patients: a meta-analysis of prospective studies. *JAMA*. 1998;279:1200–5.
2. Davies EC, Green CF, Taylor S, et al. Adverse drug reactions in hospital in-patients: a prospective analysis of 3695 patient-episodes. *PLoS ONE*. 2009;4:e4439.
3. Conney, AH. Pharmacological implications of microsomal enzyme induction. *Pharmacol Rev*. 1967;19:317–66.
4. Garrod AE. *Inborn Errors of Metabolism*. London: Oxford University Press; 1909.
5. Snyder LH. Studies in human inheritance. IX. The inheritance of taste deficiency in man. *Ohio J Sci*. 1932:436–68.
6. Hughes HB, Biehl JP, Jones AP, et al. Metabolism of isoniazid in man as related to the occurrence of peripheral neuritis. *Am Rev Tuberc*. 1954;70:266–73.

7. Drayer DE, Reidenberg MM. Clinical consequences of polymorphic acetylation of basic drugs. *Clin Pharmacol Ther.* 1977;22:251–8.
8. Clayman CB, Arnold J, Hockwald RS, et al. Toxicity of primaquine in Caucasians. *JAMA.* 1952;149:1563–8.
9. Hockwald RS, Arnold J, Clayman CB, et al. Toxicity of primaquine in Negroes. *JAMA.* 1952;149:1568–70.
10. Kalow W, Staron N. On distribution and inheritance of atypical forms of human serum cholinesterase, as indicated by dibucaine numbers. *Can J Biochem Physiol.* 1957;35:1305–20.
11. Lockridge O. Genetic variants of human serum cholinesterase influence metabolism of the muscle relaxant succinylcholine. *Pharmacol Ther.* 1990;47:35–60.
12. Kalow W, Gunn DR. Some statistical data on atypical cholinesterase of human serum. *Ann Hum Genet.* 1959;23:239–50.
13. Motulsky AG. Drug reactions, enzymes, and biochemical genetics. *JAMA.* 1957;165:835–7.
14. Vogel F. Probleme der Humangenetik. *Ergeb Inn Med Kinderheilkd.* 1959;12:65–126.
15. Kalow W. *Pharmacogenetics: Heredity and the Response to Drugs.* Philadelphia: WB Saunders; 1962.
16. Mahgoub A, Idle JR, Dring LG, et al. Polymorphic hydroxylation of debrisoquine in man. *Lancet.* 1977;2:584–6.
17. Eichelbaum M, Bertilsson L, Sawe J, et al. Polymorphic oxidation of sparteine and debriso-quine: related pharmacogenetic entities. *Clin Pharmacol Ther.* 1982;31:184–6.
18. Kalow W. Pharmacogenetics: a historical perspective. In: Thomas SJ, ed. *Pharmacogenomics: Applications to Patient Care.* Kansas City, MO: American College of Clinical Pharmacy; 2004:251–69.
19. Gonzalez FJ, Skoda RC, Kimura S, et al. Characterization of the common genetic defect in humans deficient in debrisoquine metabolism. *Nature.* 1988;331:442–6.
20. Aklillu E, Persson I, Bertilsson L, et al. Frequent distribution of ultrarapid metabolizers of debrisoquine in an Ethiopian population carrying duplicated and multiduplicated functional CYP2D6 alleles. *J Pharmacol Exp Ther.* 1996:278:441–6.
21. Latif DA, McKay AB. Pharmacogenetics and pharmacogenomics instruction in colleges and schools of pharmacy in the United States. *Am J Pharm Educ.* 2005;69:152–6.
22. Sansgiry SS, Kulkarni AS. The human genome project: assessing confidence in knowledge and training requirements for community pharmacists. *Am J Pharm Educ.* 2003;67:Article 39.

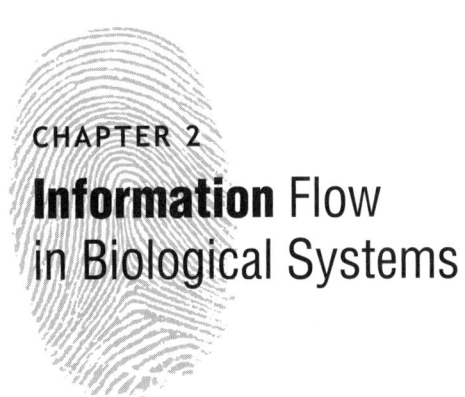

CHAPTER 2
Information Flow in Biological Systems

At the end of this chapter, you should be able to:

1. Describe, compare, and contrast the key molecules and mechanisms involved in gene expression and regulation and the subsequent synthesis of proteins in biological systems.

2. Compare, contrast, and apply simple mendelian, nonmendelian, and complex patterns of inheritance in pedigrees.

3. Identify key elements of genetic mapping studies and the precautions associated with these studies.

4. Discuss the relevance of Hardy–Weinberg principle of equilibrium in allele frequency in population genetics.

Information flow in biological systems—that is, the ultimate expression of genetic information—is largely unidirectional. A sequence of DNA in the nucleus of a cell specifies a sequence of RNA, which acts as a messenger (hence messenger RNA, or mRNA) to direct the synthesis of polypeptide chains. These polypeptide chains ultimately form the proteins that make up most of the cellular machinery (Figure 2-1). This process is strictly one way, a principle that has been referred to as the "central dogma of molecular biology." There is no mechanism among free-living organisms to convert polypeptide chains into either RNA or DNA. Similarly, neither eukaryotes nor prokaryotes have the cellular machinery to synthesize DNA based on an RNA template. However, this feat can be performed by retroviruses, using an enzyme called **reverse transcriptase**. The process of synthesizing an RNA strand based on a DNA template is called **transcription**. Retroviruses have the ability to do the "reverse," RNA to DNA. Reverse transcriptase is one of the most fundamental tools used in genomics, which is covered in Chapter 4.

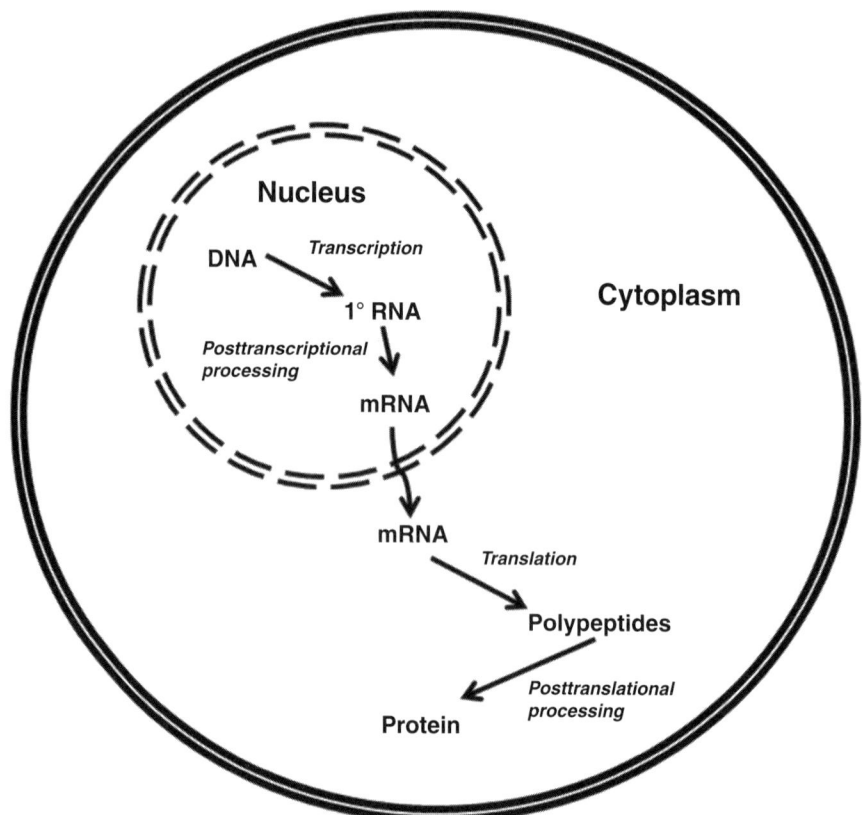

Figure 2-1. The central dogma of molecular biology. In eukaryotic organisms (those with a nucleus), the genetic instructions for the building and maintenance of the organism are in the form of DNA in the nucleus of the cell. In most cases, when a gene is "turned on," the DNA is transcribed into an initial primary RNA transcript ($1°$ RNA). Further processing results in a mature mRNA, which is then transported out of the nucleus and translated into a polypeptide sequence that ultimately undergoes further processing into the final protein.

2.1 DNA, RNA, and Proteins: The Building Blocks

Common to the structure of all DNA, RNA, and proteins are that these essential building blocks of life are composed of relatively simple, though often very large, linear polymers of a limited set of subunits. For a DNA strand or polymer, the basic subunit is a nucleotide that consists of a 5-carbon sugar (deoxyribose) co-valently bound to one of four possible **nitrogenous bases**: adenine (A), cytosine (C), guanine (G), or thymine (T). This deoxyribose sugar, with its attached base, is called a **nucleoside**. A nucleoside becomes a **nucleotide** with the addition of a phosphate group (Figure 2-2). RNA polymers are similar to DNA, with two exceptions: (1) a ribose sugar replaces deoxyribose, and (2) the base uracil (U) replaces thymine.

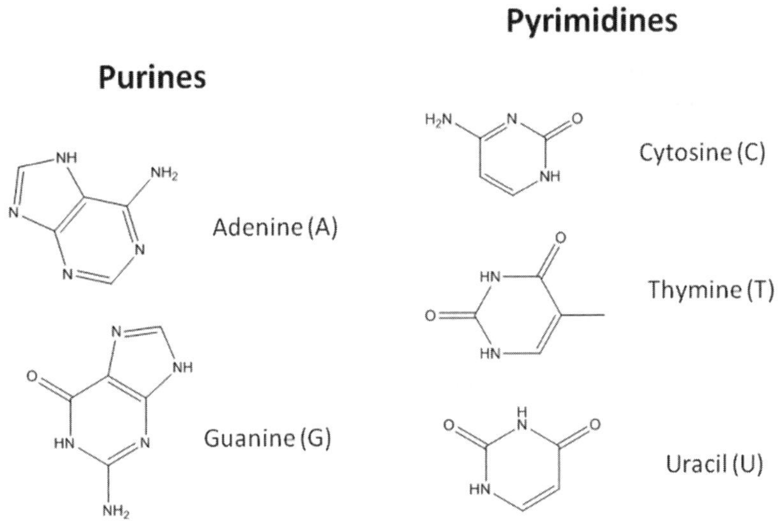

Purines

Pyrimidines

Adenine (A)

Guanine (G)

Cytosine (C)

Thymine (T)

Uracil (U)

Figure 2-2. The four nucleotides that are the building blocks of DNA. Note that in RNA, the nucleotide uracil replaces thymine.

Reverse transcriptase:
An enzyme capable of synthesizing a DNA strand using an RNA template. Retroviruses are the only known entities that have functional reverse transcriptase. Reverse transcriptase is a fundamental tool in molecular biology research for the conversion of mRNA into its complementary DNA (cDNA). A step necessary for DNA sequencing, amplification in PCR, or cloning.

Transcription:
The process of synthesizing an RNA strand based upon a DNA template. This is the first step in the expression of a gene. The enzyme responsible for the synthesis of the RNA is called RNA polymerase.

Nitrogenous bases:
For a DNA strand or polymer, the basic subunit is a nucleotide that consists of a 5-carbon sugar (deoxyribose) covalently bound to one of four possible nitrogenous bases: adenine (A), cytosine (C), guanine (G), or thymine (T).

Nucleoside:
A deoxyribose sugar, with its attached nitrogenous base, is called a *nucleoside*.

Nucleotide:
A nucleoside becomes a *nucleotide* with the addition of a phosphate group.

Watson-Crick rule:
In a DNA molecule, adenosine specifically binds to thymine via two hydrogen bonds, and cytosine binds to guanine via three hydrogen bonds.

DNA polymerase:
The enzyme that copies one DNA strand using the complementary strand as a template.

Homo/heteropolymer:
In cases where more than one polypeptide forms the protein, the polypeptides may be identical (homopolymer) or different (heteropolymer).

The structure of DNA consists of two DNA strands running antiparallel to one another, forming a double helix (Figure 2-3). The two strands are held together by weak hydrogen bonds between opposing bases, which form the internal lattices of the double helix. The base pairings are not random but follow what has come to be known as the **"Watson-Crick rule"**: Adenosine specifically binds to thymine via two hydrogen bonds, and cytosine binds to guanine via three hydrogen bonds. Given this rule, the sequence of one DNA strand can always be used to synthesize the opposing strand. The two strands of the DNA duplex are said to

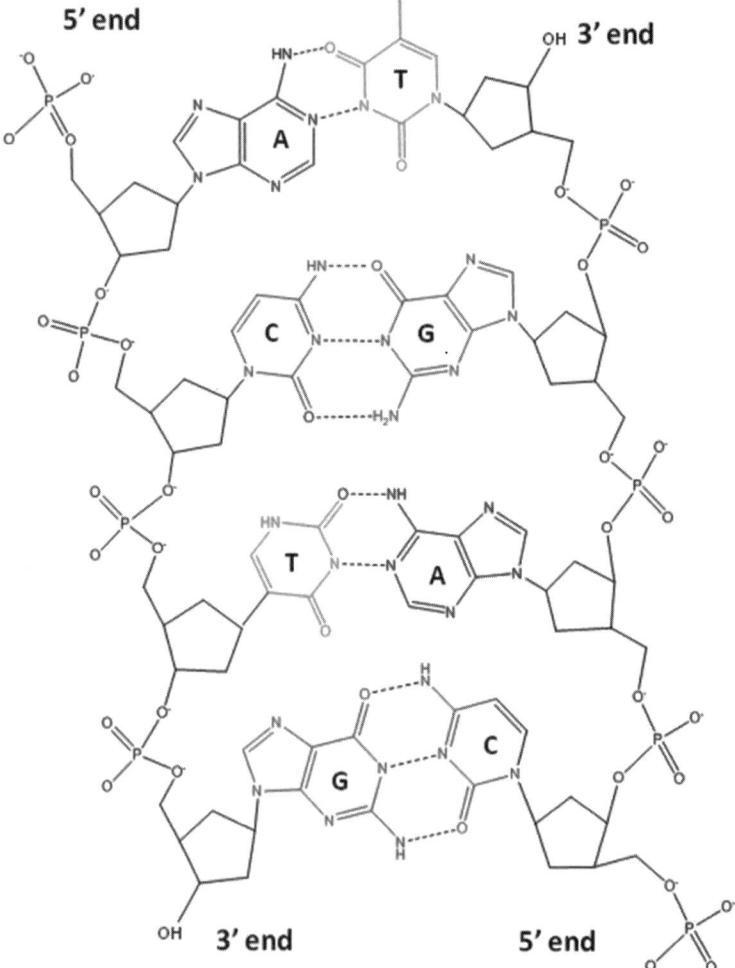

Figure 2-3. DNA, or deoxyribonucleic acid, consists of two antiparallel strands of nucleotide polymers. Adenosine (A) always bonds with thymine (T) (or uracil [U], in RNA) by two hydrogen bonds (dashed lines), and guanine (G) always bonds with cytosine (C) by three hydrogen bonds.

be complementary to one another. The discovery of DNA's structure immediately suggested a mechanism by which DNA can be copied from one cell to the daughter cell or, more importantly, from one generation to the next. The enzyme that copies one DNA strand using the other strand as a template is called **DNA polymerase** and, like reverse transcriptase, is fundamental to genomic research (see Chapter 4). RNA molecules, unlike DNA, normally exist as single strands in the cytoplasm of the cell.

Proteins are composed of one or more polypeptide strands consisting of repeating subunits of amino acids. There are 20 amino acids, all having a positively charged amino group and a negatively charged carboxyl group (carboxylic acid) joined by a central carbon. The additional side chains bound to this central carbon distinguish the 20 amino acids and give each unique biochemical characteristics. Amino acids are broadly grouped into subclasses of acidic, basic, nonpolar, or uncharged polar molecules (Table 2-1). The final structure of a protein is considerably more compli-cated than that of the DNA that encodes it. Proteins are very large macromolecules that are often composed of one or more polypeptides. In cases where more than one polypeptide forms a protein, the polypeptides may be identical (**homopolymer**) or different (**heteropolymer**). Folding of the macromolecule is determined in part by the amino acids that make up the polypeptides but also by additional alterations (for example, glycosylation) that may be specific to a given cell type.

Table 2-1. Three Basic Subclasses of Amino Acids

Subclass	Amino Acids (abbreviations)				
Nonpolar	Glycine Gly G	Alanine Ala A	Leucine Leu L	Isoleucine Ile I	Proline Pro P
	Methionine Met M	Phenylalanine Phe F	Tryptophan Trp W		
Uncharged polar	Asparagine Asn N	Glutamine Gln Q	Serine Ser S	Threonine Thr T	Tyrosine Tyr Y
	Cysteine Cys C				
Charged polar	Aspartic acid Asp D	Glutamic acid Glu E	Lysine Lys K	Arginine Arg R	Histidine His H

2.2 Gene Expression: RNA Transcription and Processing

As noted earlier, genetic information flows in one direction in all cells (DNA →
RNA → polypeptides). The linear sequence of DNA determines the sequence
of a given RNA molecule, and this RNA sequence determines the sequence of
the amino acids that make up the polypeptide (the central dogma of molecular
biology). The first step, DNA to RNA, is called *transcription* and occurs in the
nucleus of the cell, though some transcription also occurs in two special cellular
organelles, mitochondria and chloroplasts (the latter being found only in plants
and some protists), which have their own genetic material. The DNA-directed
synthesis of an RNA molecule is catalyzed by the enzyme RNA polymerase
(remember that DNA-to-DNA synthesis is accomplished via DNA polymerase).
RNA polymerase follows the same rules of Watson-Crick base pairing, except that
a uracil is added to the RNA strand wherever an adenosine occurs in the DNA
template. Many of these RNA transcripts do not go on to code for polypeptides.
These "noncoding" RNAs include ribosomal RNA (rRNA), transfer RNA (tRNA),
small nuclear RNAs (snRNA), and micro RNAs (miRNA). Those RNA transcripts
that do code for polypeptides—mRNAs—are further processed within the cell in a
series of processing steps that (Figure 2-4):

1. Remove internal sequences (introns) that do not encode the polypeptide
   chain. The encoding RNA sequences (exons) are then spliced together.
   The splice junctions delineate the boundary between the exon and the
   intron and include specific sequences (often a GU at the beginning of
   the intron and an AG at the end) that are recognized by the protein–RNA
   complex responsible for the splicing reaction (the spliceome). This RNA
   splicing step often uses different combinations of exons to produce
   different polypeptides in a process called alternative splicing. Greater
   than 50% of human genes code for slightly different polypeptides via
   alternative splicing, giving rise to protein isoforms whose function, site of
   action, or time of action vary.

2. Add or cap the five prime (5') end of the transcript with a 7-methylguano-
   sine (m^7G) triphosphate (ppp). This m^7G cap is thought to protect the
   transcript from exonuclease digestion, facilitate transport to the cytoplasm,
   and aid in the later attachment of ribosomal subunits for translation.

3. Add or cap the three prime (3') end of the transcript with a series of
   adenosines (polyadenylation) to add a polyadenosine [poly(A)] tail to the
   mature mRNA. Like the 5' end m^7G cap, the poly(A) tail on the 3' end
   of the mRNA is thought to stabilize the mRNA, facilitate transport to the
   cytoplasm, and aid in the initiation of translation.

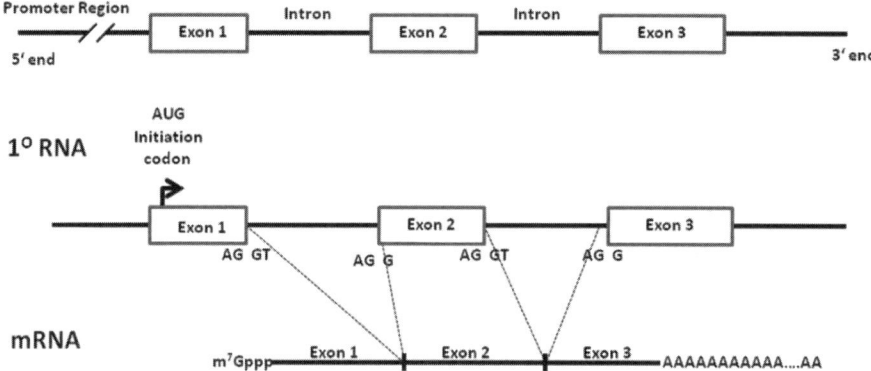

Figure 2-4. Processing steps from the DNA transcript as it would be on the chromosome to the mature mRNA. On the chromosome, a gene may be fragmented into a number of exons separated by introns, which may be many times larger than the exons. Generally upstream of the gene is a regulatory region that includes the promoter, which represents the first level of gene regulation. Upon being transcribed, the primary (1°) RNA must be further processed, including the removal of the introns and the addition of a poly[A] tail on the three prime (3') end and a 7-methylguanosine (m⁷G) cap on the 5' end. From this single primary RNA transcript, slightly different polypeptides could be assembled (for example, exons 1-2-3, or 1-3 or 2-3) in a process called *alternative splicing*.

The initiation of transcription, which is the first step in gene expression, is accomplished by the interaction of the RNA polymerase with many protein "helpers" that recognize short, specific DNA sequences in the immediate vicinity of the gene. These short sequences not only facilitate the recruitment of the RNA polymerase–protein complex but also act as binding sites for special proteins that, upon binding, regulate gene transcription or expression. These proteins, or **transcription factors**, are themselves coded for elsewhere in the genome and must travel to their binding sites. The DNA sequences recognized by the transcription factors are referred to as cis-acting. Many cis-acting elements are found on the DNA strand, generally upstream of the gene to be regulated, though there are examples of cis-acting elements many thousands of base pairs away from, or even downstream of, the transcription start site of the gene they regulate. The plasticity of these

Transcription factors:
Special proteins that, upon binding, regulate gene transcription or expression. Transcription factors are coded for by genes that are not adjacent to the genes they regulate and therefore referred to as trans-elements or factors. The sequences of DNA that bind transcription factors generally near the genes being regulated are referred to as cis-elements or cis-acting.

regulatory regions has greatly confounded our understanding of gene regulation. Major types of cis-acting elements include:

- **TATA box**—A core DNA sequence generally located 25 base pairs upstream of the transcription start site. The actual sequence is more often TATAAA or some variant of this sequence.
- **GC box**—The **consensus sequence** GGGCGG is found in many genes and is particularly associated with **housekeeping genes**. Genes lacking a TATA box often have a GC box.
- **CAAT box**—The consensus sequence GGNCAATCT occurs 75 to 80 base pairs upstream of the transcription start site.

These sequence elements occur in the **promoter region** of each gene. Generally, one or more of these specific cis-acting elements occur in every gene. In addition, more distal to this sequence there are often other specific recognition sequences that play a role in tissue-specific gene expression. These recognition sequences are very diverse and complex. Recognition sequences that increase or enhance gene expression are often referred to as **enhancers**, whereas sequences that reduce or inhibit gene expression are called silencers. In both cases, these regulatory sequences can occur thousands of base pairs away from the genes they regulate, even in introns. These sequences are recognized by specific proteins, or transcription factors, whose binding to the DNA affects gene regulation. Because these transcription factors are coded for by other genes elsewhere in the genome (not in proximity to the genes they regulate), they are often called trans-acting elements or factors (Figure 2-5). Any given gene may have a combination of multiple enhancer and silencer elements, allowing for gene expression that is both finely controlled and complex.

Consensus sequence:
A sequence that indicates the most abundant residues (e.g., nucleotides) at each position, based on multiple sequences. For nucleotide sequences, "N" stands for any base, "Y" represents any pyrimidine base, and "R" indicates any purine base.

Housekeeping genes:
Genes whose transcription rate is relatively constant with differing cellular conditions. Often the protein products of housekeeping genes are needed for typical maintenance of the cell.

Promoter region:
A DNA sequence immediately adjacent to and largely upstream of a gene. Sequences located here facilitate the binding of RNA polymerase and proteins (transcription factors) necessary for transcription.

Enhancers:
Recognition sequences that increase or enhance gene expression are often referred to as enhancers, whereas sequences that reduce or inhibit gene expression are called silencers.

Codon:
A set of three adjacent nucleotides.

Regulatory Sequence Elements	Description and transcription factors
TATA	Assembly site for RNA polymerase
EC	Enhancer core, binds C/EBP, may interact with A5 element
A1, A3, A5	Binds Pancreatic duodenum homeobox-1 (PDX-1)
GG1, GG2	GG1 also known as A2, both may also bind PDX-1
CRE1, CRE2	Cyclic AMP response elements
C2	Binds PAX4 (repressor) and PAX6 (stimulates)
E1, E2	E1 binds USF (stimulates), E2 binds IEF1
NRE	Negative Response Element, binds OCT1
G1	Binds PUR-1/MAZ

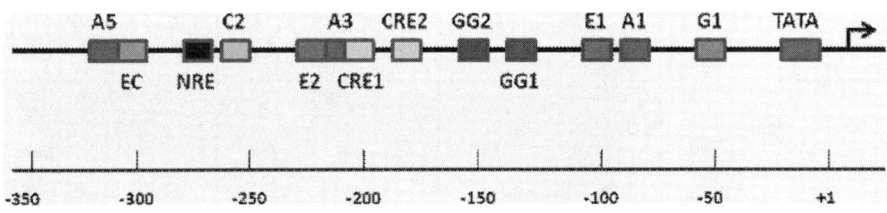

Figure 2-5. Hypothetical regulatory region upstream of a gene. The sequence positions of the cis-acting regulatory elements are noted by negative numbers, counting away from the first start codon (+1). Shown are a number of regulatory sequence elements that bind one or more transcription factors to regulate the expression of the gene. The regulation can be positive or negative, and multiple transcription factors may bind to the regulatory region simultaneously, allowing for exquisite control of gene expression.

2.3 RNA to Proteins: Translation

The now complete mRNA transcript, with the introns spliced out and with its poly(A) tail and m7Gppp cap, migrates out of the nucleus into the cell's cytoplasm, where ribosomes are enlisted to direct the synthesis of a polypeptide chain based on the mRNA sequence. Translation is based on a triplet code (the **codon**) such that each successive three nucleotides on the mRNA strand code for a specific amino acid. Each codon is recognized by a specific tRNA based on base pair complementarity. Different tRNAs carry one of the 20 different amino acids. Thus, the ribosomal complex moves down the mRNA, stopping at each three-base-pair codon long enough to recruit the appropriate tRNA with its amino acid to add to the growing polypeptide chain. Because three bases define a codon and there are four possible bases (A, C, G, and U) for each base position, the triplet genetic code has 64 different codons ($4^3 = 64$). However, there are only 20 amino acids, leaving 44 codons potentially unused. Three codons, called stop codons, are used to signify the end of the transcript, and the remainder are used over again for the 20 amino acids. Thus, each amino acid is specified on average by three codons, making the genetic code redundant (Table 2-2). One very important outcome of this redundancy in the genetic code is that some single-base-pair mutations in the original DNA strand will result in no change in the amino acid sequence of the protein. Because the amino acid sequence is unchanged, these mutations often have no effect on the organism and are said to be "silent."

Table 2-2. Specification of Amino Acids by Codons

Codon 1 & 2	Codon 3	Amino Acid	Codon 1 & 2	Codon 3	Amino Acid	Codon 1 & 2	Codon 3	Amino Acid	Codon 1 & 2	Codon 3	Amino Acid
AA	A / G	Lys	CA	A / G	Gln	GA	A / G	Glu	UA	A / G	STOP
AA	C / U	Asn	CA	C / U	His	GA	C / U	Asp	UA	C / U	Tyr
AC	A / G / C / U	Thr	CC	A / G / C / U	Pro	GC	A / G / C / U	Ala	UC	A / G / C / U	Ser
AG	A / G	Arg	CG	A / G / C / U	Arg	GG	A / G / C / U	Gly	UG	A	STOP
									UG	G	Trp
AG	C / U	Ser							UG	C / U	Cys
AU	A	Ile	CU	A / G / C / U	Leu	GU	A / G / C / U	Val	UU	A / G	Leu
AU	G	Met									
AU	C / U	Ile							UU	C / U	Phe

Legend:

A	Adenine	Purines
G	Guanine	
C	Cytosine	Pyrimidines
U	Uracil	

Abbreviations: Ala = alanine; Arg = arginine; Asn = asparagine; Asp = aspartic acid; Cys = cysteine; Gln = glutamine; Glu = glutamic acid; Gly = glycine; His = histidine; Ile = isoleucine; Leu = leucine; Lys = lysine; Met = methionine; Phe = phenylalanine; Pro = proline; Ser = serine; Thr = threonine; Trp = tryptophan; Tyr = tyrosine; Val = valine.

This is the genetic code used by most, though not all, eukaryotes to translate the mRNA strand into a polypeptide chain. Note: nucleotides are those found in RNA. The third nucleotide in the codon may change ("wobble") and yet code for the same amino acid. For example, GGA, GGG, GGC, and GGU all code for glycine.

Another type of mutational change that follows from the linear nature of translation involves insertion or deletion mutations of the DNA in coding regions. These mutations result in shifts in the reading frame of the codons (frame shift mutations). Losses or gains of nucleotides in a gene that are not multiples of three will convert an mRNA sequence into nonsense and result in loss of translation.

After translation, polypeptide chains undergo a variety of modification reactions involving the addition of chemical groups, including:

- Hydroxylation, or addition of an OH group
- Phosphorylation, or addition of a PO_4^- group
- Acetylation, or addition of a CH_3CO group
- O-glycosylation, with the addition of a complex carbohydrate

These posttranslational modifications affect the nature of subsequent folding of the final protein. Finally, the polypeptide may undergo cleavage into smaller products. In many instances, short polypeptide sequences acting as tags to direct the movement of the polypeptide to a specific location must be removed once they have served their purpose.

We have now covered information flow in a molecular sense, that is, from the long-term storage of information in DNA, to its conversion to mRNA to be transported out of the nucleus, to be ultimately converted into the working components of the cell in the form of proteins. For each of these processes—transcription, translation, and cis-acting regulatory elements—the defining characteristic is that they are all based on the linear sequence of either DNA or RNA. Changes in the sequence, or mutations, can have a variety of effects in an organism, ranging from loss of expression to changes in protein function. We now turn to information flow in a genetic sense—information flow not in a cell, but through time.

2.4 Genes in Pedigrees: Information Transmission and Inheritance

Humans are **diploid** animals. Each of us has two complete, more or less identical sets of genes arrayed on 22 chromosomes, plus a pair of sex chromosomes. Males have an X and a Y chromosome, and females have two X chromosomes. In each generation, genetic material is divided evenly into gametes through the process of **meiosis**. Each gamete receives a complete set of 23 homologous chromosomes, to be later reunited via fertilization. This is the fundamental basis of genetics in humans. Humans thus have two copies of every gene, except in cases where mutations have resulted in the loss of genetic material (**gene**

deletion). However, the genetic material in the mitochondria has a very different pattern of transmission: during fertilization, the mitochondria in sperm cells are not passed to the developing zygote. Only the egg's mitochondria become part of the zygote's genome. Thus, mothers pass their mitochondrial genome to all their offspring, both sons and daughters. The male's mitochondrial genome ends with him and is not passed to the next generation.

2.4.1 Simple Mendelian Patterns of Inheritance

Most traits are the result of the interaction of multiple genes (**polygenic**) and **environmental factors**. For example, height is certainly in part a function of genetics; tall parents generally have tall children. That the relationship between parental height and offspring height is not perfect is an indication that more than one gene is involved. In addition, the influence of environmental factors (e.g., improving diets and health care) can also be seen to influence height—on average, we are taller than our parents, and they are taller than their parents. A trait whose expression is largely determined by a single gene (**monogenic**) and is not influenced by environmental factors is said to display *mendelian inheritance*. Mendelian traits have the simplest patterns of inheritance. A classic example of a mendelian trait in humans is sickle cell anemia, which is caused by a single mutation in the beta subunit of hemoglobin. The mutation is a change of a single nucleotide in the hemoglobin-beta gene, which is located on the 11th chromosome. This single-nucleotide polymorphism (SNP) has two alleles segregating in human populations. The wild-type allele is defined by having an adenosine (A) in the sixth codon of the gene. The variant allele, which gives rise to the sickle cell trait, has a thymine (T) instead of an adenosine. The SNP results in the substitution of a hydrophilic glutamate (codon GAG) for a hydrophobic valine (codon GTG). This amino acid substitution causes aggregation of hemoglobin within red blood cells, resulting in a sickle shape. Sickle cells are less elastic than normal round red blood cells and tend to block capillaries. In addition, sickle cells are destroyed in the spleen, leading to excessive hemolysis. Individuals must carry

Diploid:
One set of chromosomes in an organism or cell containing two sets of chromosomes (paternal and maternal) is called haploid. If both sets of chromosomes are present, the case for most animal cells except the gametes, the cell is said to be diploid.

Meiosis:
Process during gamete formation in which the numbers of chromosomes per cell are divided in half. Prior to this reduction in genetic material, the DNA must be replicated.

Gene deletion:
A specific insertion–deletion mutation where the genetic sequence deleted or missing includes a functioning gene.

Environmental factors:
In genetics, "environmental factors" has a specific meaning referring to all the factors that influence the expression of a trait other than genetic factors. Environmental factors can include diet, drug, or chemical exposures, or even prenatal development.

the mutation on both chromosomes (homozygous) to exhibit the sickle cell anemia **phenotype**. Such individuals have the **genotype** TT. Because the trait is seen only in homozygous individuals, it is said to be recessive, that is, the phenotype is masked or hidden whenever the A allele is present. Heterozygous individuals (AT or TA), or carriers, may have symptoms under low-oxygen conditions. Note that the terms *dominance* and *recessiveness* simply refer to the relationship between two alleles and a defined phenotype. A dominant allele need not be the most common allele in a population, and its frequency in a population will not increase in time due to dominance.

For mendelian traits displaying strict dominance–recessiveness, there are five basic patterns of inheritance, depending upon whether the gene responsible for the trait is autosomal or located on one of the sex chromosomes (Figure 2-6). Most pedigrees, like the ones shown in Figure 2-6, display the phenotypic status of each individual by showing symbols that are either filled (affected individuals) or open (unaffected individuals) and square (male) or round (female). Not shown, though it may be possible to discern from the patterns of inheritance, is each individual's genotype. For example, in Figure 2-6, the female I-2 in the first pedigree (whose genotype is autosomal dominant) displays the condition (filled circle), and based on the pattern of the trait in her children, she

Monogenic:
A trait whose expression is largely determined by a single gene. Traits of phenotypes affected by multiple genes are said to be **polygenic**.

Phenotype:
The observable outcome of the interaction of an individual's genes and environmental factors.

Genotype:
The underlying genetic constitution of an individual, usually in relation to a specific trait. **Genotyping** refers to tests carried out to determine an individual's genotype.

Hemizygous:
Cases in diploid cells where there is only one copy of a chromosome or chromosomal region. Males having only one X chromosome are said to be hemizygous for the sex chromosomes.

is likely to be heterozygous. If she were homozygous, we would expect to see all her children affected, because the trait is dominant. Traits caused by genes on the autosomes affect both sexes equally. Traits caused by X-linked genes are most often seen in males because males are **hemizygous**—that is, they have only one X chromosome. Therefore, any X-linked trait, whether dominant or recessive, will always present in males, because they have only one copy of that gene. For X-linked recessive traits, females will be affected only if they are homozygous, which is unlikely if the causative allele is rare. For example, if an X-linked, recessive mutation has an allele frequency of 0.001 (i.e., found on 1 of every 1,000 chromosomes), the frequency of affected males will be 0.0001 and the frequency of affected females will be 1×10^{-6} (0.001 × 0.001, or 1 in 1 million). Y-linked traits do not affect females because they have no Y chromosome. A father with a Y-linked trait will pass that trait to all his sons, all his grandsons, and so on. To date, the only

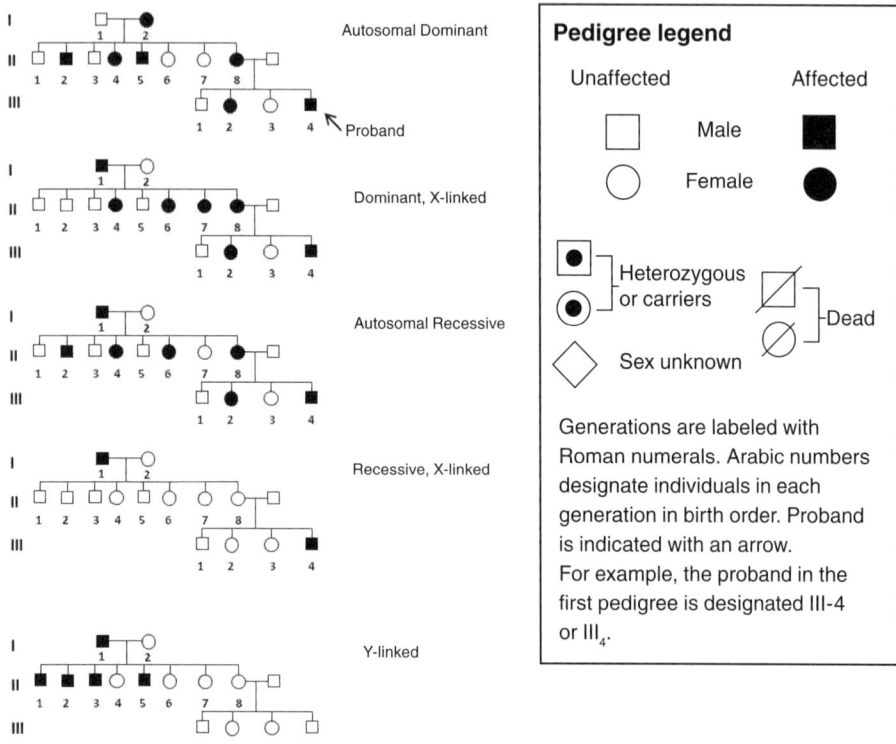

Figure 2-6. Pedigrees for the five basic types of mendelian inheritance patterns. For these pedigrees, as is often the case, squares represent males, circles represent females, and the filled symbols indicate that the individual displays the phenotype of interest. The genotype is not indicated. For example, for a dominant trait, an individual will exhibit the trait (filled symbol), whether homozygous or heterozygous. The arrow indicates the proband, or the individual who is the subject of study. As an exercise, the reader is encouraged to determine the genotypes of each of the individuals in the five pedigrees.

traits that have been conclusively shown to reside on the Y chromosome are genes for male differentiation and spermatogenesis. One would not expect to find genes that are important for both sexes on a Y chromosome, because females, lacking a Y chromosome, would not have that gene.

A common departure from simple mendelian patterns of inheritance occurs when the products of both alleles are detectable in the individual. Such traits are said to be **co-dominant**. One of the earliest known examples of co-dominance in humans is MN blood types. Here a single gene is responsible for antigenic specificities, either M or N. Heterozygotes produce both the M and N antigens and can be detected. Slightly different situations can arise when the heterozygote expresses with an intermediate phenotype rather than with both alleles. These cases of **semi-dominance** may arise when one of the alleles of a gene results in loss of expression. Heterozygotes would potentially produce half the gene product, which may be detectable as reduced function. Although such mutations are fairly common, examples of semi-dominance, particularly clinically important cases, are actually rare because living systems are robust. For most genes, having one functioning copy or allele is sufficient for normal function. Heterozygous individuals with one normal allele and one **null allele** are, with few exceptions, phenotypically normal. This is simply another way of describing a recessive trait. Most null alleles are recessive in that the loss-of-function allele is masked whenever a normal allele is present. Dominance and recessiveness are the straightforward outcome of robust living systems.

Next in increasing complexity are traits whose phenotypic expression is due to two or a few genes. Consider coat color in Labrador Retrievers, whose coat colors are black, chocolate, or yellow and are determined by two genes. One of the genes determines the pigment, either black or brown, to be deposited in hair follicles. Black is dominant to brown. The other gene determines whether the pigment is deposited in the hair follicles. The dominant allele results in pigment deposition and either black or brown Labrador Retrievers. The recessive allele results in no deposition in the hair follicles and the dogs are yellow, independent of the alleles at the other gene (Figure 2-7).

Co-dominant:
A common departure from simple mendelian patterns of inheritance occurs when the products of both alleles are detectable in the individual.

Semi-dominance:
Pattern of inheritance where the heterozygote has an intermediate phenotype. For example, cases of semi-dominance may arise when one of the alleles of a gene results in loss of expression. Heterozygotes would potentially produce half the gene product, which may be detectable as reduced function.

Null allele:
A variant allele that produces no gene product.

Epistasis:
A situation in which two or more genes interact in a nonadditive way to produce a phenotype.

Epistasis refers to situations in which two or more genes interact in a sometimes nonadditive way to produce a phenotype. In many cases, phenotypes that are largely caused by a single gene are often modified by other genes in the genome. For example, cystic fibrosis is an autosomal recessive disorder caused by mutations in the cystic fibrosis transmembrane conductance regulator gene. However, patients with cystic fibrosis show significant phenotypic variation, probably due to other alleles on other genes. As is the case for many human traits, environmental factors also modify the phenotype.

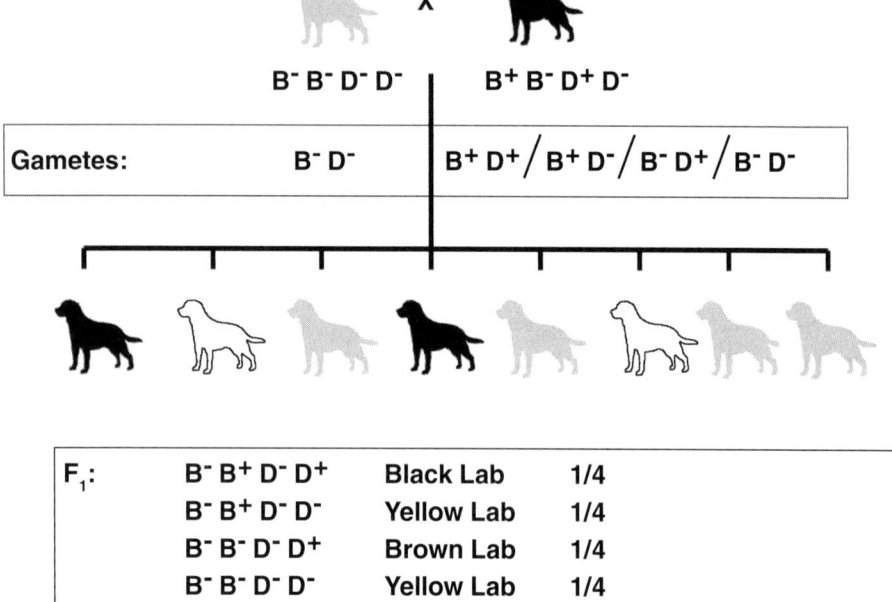

F₁:			
	B⁻ B⁺ D⁻ D⁺	Black Lab	1/4
	B⁻ B⁺ D⁻ D⁻	Yellow Lab	1/4
	B⁻ B⁻ D⁻ D⁺	Brown Lab	1/4
	B⁻ B⁻ D⁻ D⁻	Yellow Lab	1/4

Figure 2-7. A classic example of epistasis between two genes that determine coat color in Labrador Retrievers. One gene (B) is responsible for pigmentation, either brown, or chocolate (B⁻ recessive); or black (B⁺ dominant). A second gene determines whether the pigment is deposited in the coat (D⁺ dominant) or not (D⁻ recessive). A Labrador Retriever who carries the recessive gene for deposition (D⁻D⁻) on both chromosomes will be yellow, regardless of the genotype at the pigmentation locus.

2.4.2 Complex Patterns of Inheritance

Most of the well-known genetic disorders, such as sickle cell anemia, Huntington disease, Duchenne muscular dystrophy, and cystic fibrosis, are familiar to us precisely because they are single-gene, or monogenic, disorders. Unfortunately, most traits and diseases that afflict humankind are not caused by a single gene, and therefore the genetic basis for these traits is not so easily understood. Given that the human genome has more than 20,000 genes, it shouldn't be surprising that most human traits are the result of the interaction of multiple genes and environmental factors. These traits display complex patterns of non-mendelian, or multifactorial inheritance, and their study is the basis of the field of quantitative genetics.

As the name suggests, the relationship between genotype and phenotype is much less clear for complex traits. Although individual alleles follow Mendel's law of independent assortment and segregation, the distribution of phenotypes becomes continuous rather than discrete as the number of genes playing a role in the traits' expression increase (Figure 2-8). Most human traits, such as height, intelligence, susceptibility to disease, and drug response, exhibit variation much like that shown in Figure 2-8B. The goal of quantitative genetics is to partition the variability seen in a given trait into those portions that are due to genetic factors and those due to environmental factors.

Perhaps one of the most straightforward and commonly used methods to assess the genetic component that contributes to phenotypic variation is the study of twins. Monozygotic (identical) twins have the same DNA, and therefore any differences in a trait of interest are due to environmental influences or factors. For the same reason, studies of families are also valuable in assessing the **heritability** of a phenotypic trait. A trait with high heritability is one in which a large portion of the

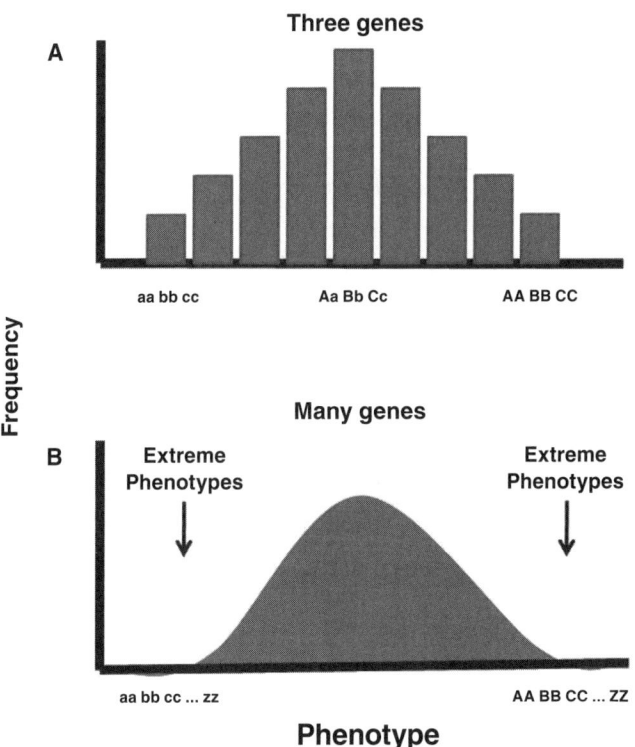

Figure 2-8. (A) Frequency histogram for a trait controlled by three genes (A, B, C). In this example, each gene has an equivalent and additive effect on the trait. This is almost never the case. Oftentimes there are a few genes with large effects and many genes with small effects. (B) Continuous distribution of a trait when multiple genes are involved. Often in genetic mapping studies, individuals exhibiting extreme phenotypes are chosen to participate with the assumption that they carry more of the mutations than individuals whose phenotype is intermediate.

variation in the trait is due to genetics. Conversely, a trait with low heritability is one in which most of the variation in the trait among individuals is due to environmental causes. Plant and animal breeders hope that the traits they value (e.g., milk production, fruit yields) have high heritabilities. If heritabilities are low, there is very little for breeding programs to select for, and it is better to improve environmental conditions.

2.5 Genetic Mapping Studies

If the trait of interest has a heritable component, the challenge is then to identify the genes involved. Here again, the best studies involve individuals who are related, for example, family studies. Related individuals have similar **genetic backgrounds**; that is, they share many alleles at their genes. Having similar genetic backgrounds among study subjects aids in genetic mapping studies, because if multiple genes are involved to produce a phenotype, these individuals share many of the same "players" (alleles). In addition, family members often are exposed to many of the same environmental conditions (e.g., diet, health

Heritability:
The proportion of total phenotypic variance due to genetic variation.

Genetic background:
The totality of all the genes within an individual. Given gene–gene interactions, the expression of any allele will be affected by the other alleles processed by an individual. The expression of an allele may change if the other alleles with which it interacts are different.

Genetic marker:
A region of DNA of known location in the genome that is polymorphic (has allelic differences). These alleles could be SNPs or indels in the DNA.

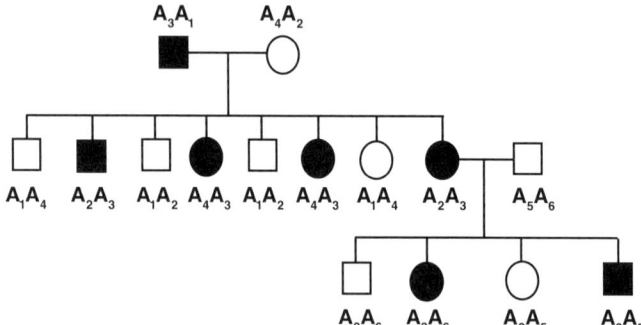

Figure 2-9. Pedigree showing genotypes for all individuals of one family. Such information, both genotype and phenotype, is necessary for linkage studies. Studies would include pedigrees for many families to assess linkage of a trait with a specific allele. In this example, all the family members have been genotyped at one locus (A), and there are six alleles (A1 through A6). Based upon this single family, one would hypothesize that the trait may be linked to one of the alleles (A3). Note that every time the trait appears in this family (filled symbols), that individual has at least one A3 allele; also, there are no unaffected individuals with the A3 allele. Thus the A3 allele is in linkage disequilibrium with the trait—they are not assorting independently, implicating the region around A3 as harboring the causative gene.

care), thus reducing the contribution of environmental factors to the overall variance in the expression of the trait. The goal of these studies is to detect a significant correlation between a **genetic marker** and the trait of interest. Often a chromosomal region is evaluated for the presence of genes that may affect the trait. Individuals—both affected and normal (controls)—are then genetically tested (genotyped) for each of the alleles at a battery of genetic markers in the chromosomal region. Significant correlations between one of the alleles for one of the genetic markers and the trait may indicate that the marker is close (within 100 kilobases) to an involved gene (Figure 2-9). The correlation between an allele and the trait occurs because the two loci are in **linkage disequilibrium**—they are physically close enough to one another on the chromosome that they do not segregate independently of one another, but are linked. Over many generations, recombination (during meiosis) will erode this linkage, but the closer two loci are to one another, the more time (generations) it will take to reduce the correlation between the two loci until it matches background rates. Identifying these regions of linkage disequilibrium across the chromosomes is the goal of the human **HapMap Project**.

Family studies are often difficult to carry out because very large sample sizes (i.e., large numbers of affected families) are needed when the contributions of any single gene are small. More often, researchers conduct large studies with unrelated individuals and

Linkage disequilibrium: Two genetic loci are said to be in linkage disequilibrium if they co-occur more frequently than would be expected given mendelian independent assortment. Two loci that are physically near one another on the chromosome are likely to be transmitted together; they are linked.

HapMap Project: An international effort to identify and map the haplotype blocks of the human genome. Haplotype blocks are large segments of chromosomes whose sequence is nearly identical to that of many members of a population. The aim of the HapMap effort is to describe common patterns of human variation that arise due to linkage. In this way, large blocks of the genome (thousands of kilobases) can be characterized by a limited number of diagnostic SNPs (tag SNPs).

seek to identify associations between allelic states of many genetic markers and loci that may be causative agents for the trait. Like family-based pedigree analyses, the basic principle behind association studies (often called **genome-wide association studies**) is linkage disequilibrium between random genetic markers and genes that play a role in the trait of interest. The chromosomal regions to be evaluated for the possible gene of interest can be quite large (genome-wide), limited only by the availability of genetic markers and the cost of genotyping individuals for these thousands of markers. However, a number of special precautions are necessary to ensure that the loci identified are truly involved in the trait and not false positives. These precautions are essential for the success of an association study and include:

■ ***Environmental causes***—As noted earlier, environmental factors cannot only modify the expression of traits but may also be largely responsible for the trait. For example, many cancers arise entirely due to environmental exposures (sporadic cancers). In the same sense, individuals may respond poorly to a drug because of dietary factors, other illnesses, or concomitant drugs. All are environmental factors. In conducting genetic mapping studies, it is critical to exclude those individuals whose poor response is due to environmental factors, because their inclusion will only mask any real genetic causes. Such traits that are caused by environmental factors and are indistinct from the trait being examined are called **phenocopies**. Well-thought-out exclusion criteria are essential in a mapping study to reduce the likelihood of phenocopies.

■ ***Locus heterogeneity***—When the disorder or trait of interest, like poor drug response, is caused independently by mutations in genes at different chromosomal loci, there is *locus heterogeneity*. This is different from epistasis, in which two or more genes interact to cause a trait. Locus heterogeneity arises when two or more genes alone cause the same trait. In large genetic mapping studies, like association studies, one mechanism used to minimize locus heterogeneity is to choose a study population with a homogeneous genetic background. Affected individuals with similar genetic constitutions are more likely to have the same series of genetic mutations. A related phenomenon, **allelic heterogeneity**, is very common and results when two or more mutations in the same gene give rise to the same phenotype. As described in Chapter 5, allelic heterogeneity severely limits the universality of genetic tests that assay for only one mutation in a gene as diagnostic tools to predict drug response. Other mutations in the same gene, not in the assay, may result in the same malady.

■ ***Population structure or stratification***—Correlations among a given trait and gene loci may arise when individuals within the study share haplotypes due to a common ancestral heritage. An undetected population structure within a study population may thus give rise to spurious correlations due to ancestral origins, rather than actual causative or risk-associated genes. This is particularly a problem when the underlying population structure is not the same between cases and controls in a study. Well-designed genetic mapping studies pay special attention to ensure that individuals in case and control groups are appropriately matched. In fact, most successful studies take great pains to control for the genetic background of the participants, because individuals who share the same ancestral origins have similar genetic backgrounds. This is the principal reason that many genetic mapping studies are conducted with homogeneous, often isolated populations. For example, Icelanders and Plain people (e.g., Amish or Mennonite) are often utilized in genetic studies because these populations are relatively genetically homogeneous.[1] An additional advantage among these groups is that their environmental experiences (e.g., diet, environmental exposures, health care) are often similar as well.

■ *Epistasis*—The epistatic interactions of many genes in the expression of trait under study poses severe difficulties in identifying the underlying genetic loci, particularly if there are many loci with small, nonadditive effects on the trait. The selection of individuals who exhibit the most severe forms (extreme phenotypes) of the trait in question is a mechanism commonly used to control for epistatic interactions.[2] The underlying assumption in selecting for extreme phenotypes is that if multiple interacting genes are causing the phenotype, then those individuals with the most severe phenotypes are likely to carry many of the same risk-causing alleles at most of the loci.

Genome-wide association studies (GWAS): Genetic studies that examine the entire genome for correlations among a very large number of known genetic markers and specific phenotypes.

Phenocopies: Traits that are caused by environmental factors and are indistinct from the trait being examined for genetic causes.

Allelic heterogeneity: Case where a given phenotype is caused independently by different alleles within the same gene. When the same phenotype is caused independently by two or more distinct genes, it is called locus heterogeneity.

Gene pool: All of the alleles for a given gene in a population make up the gene pool, with each allele having a specific allele frequency.

2.6 Genes in Populations: Population Genetics

Although any single individual within a population has only two alleles, there may be many more alleles segregating in that population. All of the alleles in a population make up the **gene pool**, and each allele has a specific allele frequency. Consider a gene for which there are two alleles, A_1 and A_2, with respective allele frequencies p and q. Because there are only two alleles, $p + q = 1$, the probability of drawing two p alleles is p^2. Similarly, the probability of drawing two q alleles is q^2. The probability of drawing a p and a q is pq + qp, or $2pq$. In a human gene pool, these same calculations would hold true for a gene with two alleles, assuming that individuals are chosen randomly from the population. Thus it is possible to calculate genotype frequencies on the basis of allele frequencies, as long as individuals are chosen randomly from a single gene pool. Of more use is the independent demonstration by two scientists, G. H. Hardy and Wilhelm Weinberg, that these allele and genotype frequencies will remain constant from one generation to the next if four basic assumptions are met:

1. Populations consist of large, random mating units.
2. There are no net mutation differences from one allele to another.
3. There is no differential migration of individuals into or out of the population.
4. Selection does not favor one allele over another.

Rare allele effect:
A theory to explain how recessive deleterious, even lethal alleles can be maintained in a population in spite of strong selection to remove them. As selection acts to remove a deleterious allele from a population, the frequency of the allele becomes even rarer. Because selection can act only on the homozygous individuals for a recessive allele, the opportunity to remove the deleterious allele is exceedingly rare. At the same time the deleterious allele is hidden from selection in the heterozygotes.

Populations that meet these assumptions are said to be in Hardy–Weinberg (H-W) equilibrium. This principle has two important consequences: The H-W equilibrium is very robust, and most populations are found to be in equilibrium. Therefore, when a population is found not to be in H-W equilibrium, it is usually an indication that either the genotyping method is in error (which is most often the case) or that one of the above four assumptions has been violated. If the genotyping system is trustworthy, then the next best guess is that either the population being studied is not a single, randomly interbreeding unit (assumption 1) or selection is at work (assumption 4). The other important consequence of H-W equilibrium is that, given the allele frequencies, it is relatively simple to calculate genotype frequencies for any number of alleles. Finally, an examination of allelic frequencies provides an explanation for the occurrence of deleterious alleles in a population even in the face of strong selection (**rare allele effect**): as a recessive, deleterious allele becomes rare in a population, the frequency of the recessive homozygote becomes even rarer, and thus selection has very little to act upon.

QUESTIONS

1. Certain recessive genes cause profound hereditary deafness, and individuals who are homozygous for such genes are occasionally found in high frequencies among extended families in small, isolated communities. The mutations originate in individuals from several generations in the past and become homozygous through marriages among relatives. A deaf man and a deaf woman from two different communities, each having deaf parents, have three children, all of whom have normal hearing. How would you explain this?

2. A color-blind woman marries a man with normal vision. She has a daughter who, she is glad to find, has normal vision. Her daughter marries a man with normal vision. What should she tell her daughter about the probability of having color-blind children?

3. Explain what is meant by the following statement: "The genetic background in which a mutation finds itself will have an impact on phenotype."

4. Complete the pedigree (second and third generations) below if the trait being analyzed is X-linked dominant.

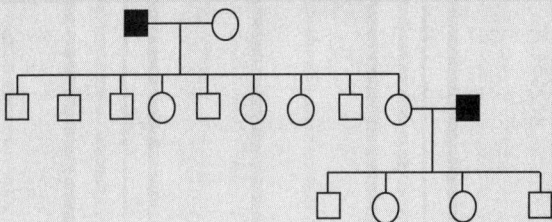

5. Why would redefining a clinical phenotype to the most extreme case help in a genetic mapping study?

6. Describe the difference between locus heterogeneity and epistasis.

7. "Tongue rolling" is determined by a single gene with two alleles:

    "T" allele → ability to roll tongue; dominant; allele frequency = 0.447
    "t" allele → inability to roll tongue; recessive; allele frequency = 0.553
    If this gene were not autosomal, but was instead located on the X chromosome:

    a. What is the frequency of males who can roll their tongues?
    b. What is the frequency of females who cannot roll their tongues?
    c. If there were 100 individuals (50 males and 50 females) in a given
        group, how many would be able to roll their tongues?

8. The mutation that causes Tay-Sachs disease, an autosomal recessive,
    neurodegenerative disease that leads to death usually before the age
    of 5 years, has remained in human populations for many hundreds of
    generations. Explain why this mutation persists in human populations
    despite its huge selective disadvantage.

9. "Most single-locus genetic mutations are recessive." Describe what is
    meant by this statement, and provide an explanation of why mutations
    are generally recessive.

10. Mutations of clinical importance can occur in both trans-acting factors
    and in cis-elements. Of the two, which is more likely to exhibit pleio-
    tropic effects?

References

1. Strauss KA, Puffenberger EG. Genetics, medicine, and the Plain people. *Annu Rev Genomics Hum Genet.* 2009;10:513–36.
2. Nebert DW. 2000. Extreme discordant phenotype methodology: an intuitive approach to clinical pharmacogenetics. *Eur J Pharmacol.* 2000;410:107–20.

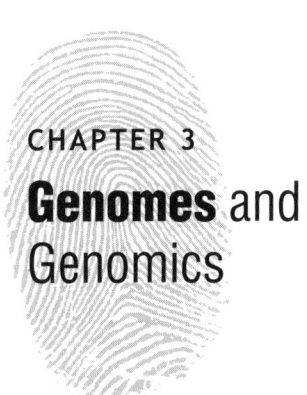

CHAPTER 3

Genomes and
Genomics

LEARNING OUTCOMES:

At the end of this chapter, you should be able to:

1. Compare and contrast:

   - Genomic and mitochondrial DNA
   - Genes and pseudogenes
   - Single-nucleotide polymorphisms and indels
   - Locus heterogeneity and allelic heterogeneity
   - Polygenic and monogenic traits
   - Homologous and orthologous genes

2. Discuss the different types of mutational events, specifically genome duplication versus subgenomic duplication; how these processes contribute to genetic diversity in individuals; and the relevance of these mutational events to important gene families that are critical to drug disposition.

3. Discuss the organization of the human genome as it relates to chromosomal structure and how various variants and alleles are named in current scientific and clinical practice.

The genome of an organism comprises all the DNA to be found in a normal cell, including both the nuclear genome, which is packaged in one or more chromosomes, and the mitochondrial genome. Within this genomic DNA (gDNA) are the genes that function to both build and maintain the organism. In most organisms, approximately 90% to 95% of the genes are transcribed into messenger RNA (mRNA) and ultimately are translated into proteins. Some genes, generally 5% to 10%, are "RNA genes," whose final products are RNA molecules that often combine with proteins (encoded by other genes) to build an RNA–protein complex.

These include ribosomal genes, transfer RNAs (tRNAs), small nuclear RNAs (snRNAs), and microRNAs (miRNAs).

Genomes are not static. As the genome sequences of many different organisms are completed, it has become evident that the content and structure of genomes are quite plastic through time. Genomes change and evolve through time. This plasticity has resulted in very complex genomes in most **eukaryotes** and many large, diverse gene families interspersed among even larger regions of noncoding DNA. The complexity of most genomes not only complicates the discovery of genes but also confounds the understanding of gene function, including the role that specific genes play in drug response. Given that real progress in understanding the nature of human disease and drug response will come only with an understanding of the intricacies of the interactions of genes with other genes and with the environment, an understanding of the nature of genomes is necessary.

3.1 Genomics

Genomics is the study of the:

- ▪ Structure of genomes—How are genomes organized, and is the organization of genes on a chromosome relevant to gene regulation and function?
- ▪ Content of genomes—What are the genes that make up a given organism's genome? Is there a relationship to the genes present in the genome and the organism's way of life?
- ▪ Evolution of genomes—How have genomes changed across a lineage, for example, that of primates? Can this knowledge reveal something about gene function?

Genome size and complexity vary greatly among organisms (Table 3-1). The first free-living organism to have its genome sequenced, *Haemophilus influenzae* (at 1.8 million base pairs), was completed in 1995. The rough draft of the human genome was completed in 2001,[1,2] and a nearly complete genome sequence was published in 2004.[3] As of 2010, well over 2,000 organisms have had their genomes sequenced, including viruses, bacteria, protists, fungi, plants, birds, amphibians, insects, fishes, and mammals. As these sequences are analyzed and compared, it has become apparent that genomes are not static and that understanding their evolution can yield much information about the evolution of all of the biochemical systems that are necessary for life. There have also been many surprises that have forced scientists to rethink the relationships between our genes and their ultimate expression in building and maintaining a human.

Table 3-1. Genome Statistics for a Number of Scientifically Important Model Organisms

Organism	Genome Size (MB)	Year Genome Sequencing Completed	No. of Chromo-somes	No. of Genes	No. of Gene Families	% Coding DNA
Escherichia coli	4.6	1997	1	4,288	2,500	~100
Yeast	14	1996	32	6,200	4,500	~20
Fruit fly	165	2000	8	13,600	11,434	~20
Rat	275	2004	40	25,557	9,969	~3–5
Human	3,000	2001	46	23,000	10,349	~3–5

Although the simplest and smallest genomes are found among eubacteria, there does not seem to be a direct relationship between genome size and organism complexity. In general, the relationship between genome size and organism complexity is evident in the genomes of **prokaryotes** but becomes far less straightforward among eukaryotes.[4] For example, most salamander and flatworm genomes are larger than mammalian genomes, and some of the largest genomes are found among the protozoa and algae. Part of this lack of correspondence is explained by the fact that eukaryote genomes contain far greater amounts of non–protein-coding DNA in regions between (intergenic) and within (intragenic or intronic) genes. Initially, this noncoding DNA was referred to as "junk" DNA, but sequence similarities of this "junk" DNA among organisms suggest that there may be some as-yet-unknown function for at least some of this DNA.

Eukaryote:
Organisms whose cells contain membrane-bound organelles including the nucleus, mitochondria, Golgi apparatus, and chloroplasts in plants. Organisms whose cells lack membrane-bound structures are referred to as **prokaryotes.**

Among eukaryotes, the genome is organized into two parts—the very large nuclear genome and the much smaller mitochondrial genome. The mitochondrial genome consists of a circular double-stranded DNA molecule. As the name suggests, the mitochondrial genome is enclosed in the mitochondrion, a membrane-bound organelle that is distinct from the nucleus. The number of mitochondria in a cell varies among organisms and among tissue types. Cells typically contain from less than 10 to well over 1,000 mitochondria, though some cells have no mitochondria at all. The mitochondrial genome is compact, lacks introns, and contains genes coding for RNA genes (in humans, 22 tRNA and 2 ribosomal RNA [rRNA] genes in humans) and polypeptides involved in oxidative phosphorylation—the production of adenosine triphosphate (ATP) (13 genes in humans). Although not all the genes necessary for ATP production are found in the mitochondria, many are found within the nuclear genome. As noted in Chapter 2, inheritance of the mitochondrial genome is very different from that of the nuclear genome in that the mitochondria within a sperm cell do not pass into the egg during fertilization. Thus the mitochondria present in the fertilized egg come only from the mother, and their inheritance is maternal. For example, Leber hereditary optic neuropathy is characterized by bilateral subacute visual failure in young adults.

> **Indel:**
> Mutations defined by either the Insertion or Deletion of nucleotides in the genome. Indels can range in size from 1 nucleotide to millions of nucleotides. Includes copy number variants and gene deletions.

The condition is caused largely by genes located in mitochondrial DNA and thus is transmitted to future generations only through mothers, not fathers. The genetic code used by the mitochondria to encode polypeptides is slightly different from that used by nuclear genomes, perhaps reflecting the independent origin of mitochondria prior to the evolution of complex, eukaryotic cells. This explains in part the inclusion of tRNA genes in mitochondrion genomes, because these tRNAs need to match the slightly different genetic code used by the mitochondria.

The nuclear genomes of organisms are arranged into much larger DNA molecules—chromosomes. Many, though certainly not all, prokaryotes have a single, circular chromosome. In addition, most prokaryotes also contain much smaller extrachromosomal DNA molecules, also circular, called plasmids. In most eukaryotes, though again not all, the chromosomes are linear and enclosed within the nucleus. Chromosomes among organisms range in size from 10,000 base pairs to over 1 million base pairs.

The ultimate source of the great genetic diversity observed not only among species, but also among individuals within species, arises from copying errors during DNA replication. With each generation, the genetic material contained in the chromosomes must be copied and then divided equally into a complete haploid set in the gametes during the process of meiosis. It is during this copying process of DNA replication that mutations can occur in the DNA sequence. These changes ultimately give rise to the diversity of life. Mutations can encompass changes from a single base pair substitution or change (single-nucleotide polymorphism [SNP]) to the addition or loss of DNA. Insertions or deletions of DNA (**indels**) can consist of a single nucleotide up to large chromosomal regions or entire chromosomes. In the context of genome evolution, that is, the building and altering of genomes through time, the mutational events we are most concerned with are relatively large and of two basic types. These large mutational events are important in pharmacogenomics because it is these events that give rise to new genes that make up gene families.

Duplications of the whole genome
- Rare
- May have been two rounds in vertebrate evolution

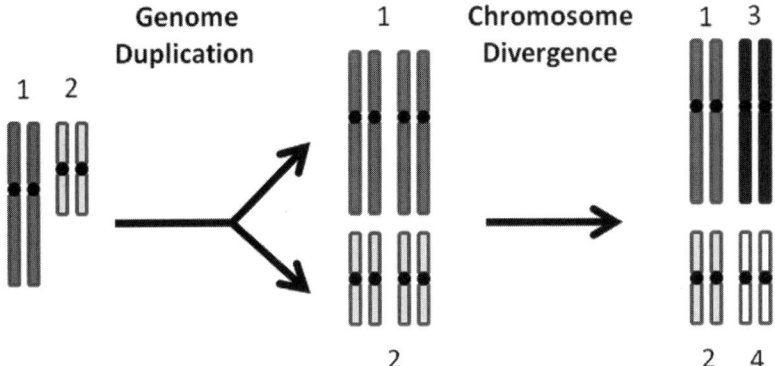

Figure 3-1. Genome building: duplication of the entire genome. Such an event, resulting in the doubling of the entire genome, is thought to be exceedingly rare. With the now doubled genetic material, the chromosomes can diverge from one another over long periods of time.

The most extreme mutational event is the duplication of the entire genome (Figure 3-1). As might be imagined, these are very rare events. **Whole-genome duplications** are thought to have occurred twice in the ancestral line, giving rise to all vertebrates,[5] and have not occurred since. Evidence for whole-genome duplications in vertebrates comes from the observation that many important single-gene clusters in invertebrates have three to four copies in vertebrates. For example, humans have four basic *Hox* gene clusters, whereas the fruit fly (*Drosophila*) has only one. Amphioxus (lancelet), a primitive chordate from a group distantly related to vertebrates, also has only one copy of this gene cluster. This doubling of genome size creates much new genetic material upon which selection may act. The mechanism for whole-genome duplication events is unknown, as is the evolutionary importance of these large-scale genomic events. Such a mechanism of genome change is no longer possible for animals that reproduce entirely by sexual reproduction, which necessitates the recombining of gametes with only the appropriate number of chromosomes. Gametes with extra chromosomes are almost always not viable.

Whole-genome duplication:
Rare mutational event in which the chromosome number is doubled or fails to be reduced during cell division. Much more common are subgenomic duplication events where smaller chromosomal regions are duplicated. Often these smaller regions contain one or more genes.

Pseudogenes:
Nonfunctional genes in the genome that are no longer expressed in any member of the species.

Subgenomic duplication events
- May result in gene & exon duplication
  - Unequal crossover
  - Unequal sister chromatid exchange
  - Interchromosomal exchange
  - Retrotransposition
  - Translocation
  - Large-scale transposition

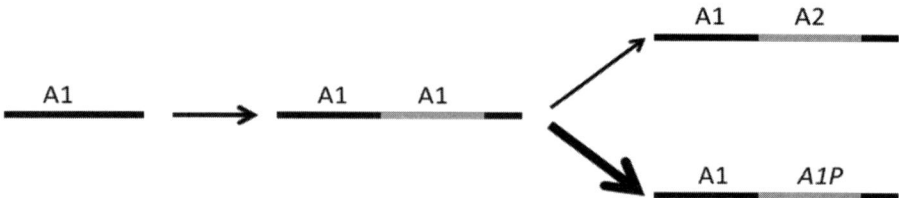

Figure 3-2. Genome building: subgenomic duplication events. Much more frequent than whole-genome duplication events is the duplication of smaller regions of chromosomes. Such mutations may involve hundreds to millions of nucleotides and frequently occur during meiosis. Genes that may occur within the duplicated regions may diverge from one another through time. In most cases this divergence in sequence results in the inactivation of the gene, a pseudogene (A1P). Rarely, the sequence divergence may result in a "new" gene whose sequence codes for a "new" protein with slightly different capabilities (A2).

Another mechanism for genome change has been fundamental to genome evolution and occurs every generation and among all individuals. Subgenomic duplication events result in the duplication of chromosomal regions ranging from a few base pairs to millions of base pairs—large enough to include entire genes (Figure 3-2). Whether the region being duplicated is a large chromosomal segment or a few thousand base pairs, the fate of this "new" DNA is roughly the same. Genes contained in the duplicated regions ultimately acquire additional mutations through time. These mutations almost always render these duplicated genes nonfunctional, giving rise to **pseudogenes**—nonfunctioning "genes" whose sequences are similar to those of functioning genes. Pseudogenes represent the remnants of past duplication events. Much more rarely, a duplicated gene may acquire mutations that confer new functions. For example, through mutations, duplicated genes may acquire the ability to recognize and bind to new substrates, or to metabolize or transport new compounds. In this way, gene families consisting of "related" genes are created through time. In this context, "related" indicates that these genes have similar sequences and functions, though such sequences are also "related" in the sense that they share a common ancestral sequence.

This continual process of subgenomic duplication means that genomes are not static, but constantly change through time. In this way, the membership and size of gene families change through time, resulting in modifications in the nature of the functional scope or complexity of organisms. Gene families can both increase and decrease through time, presumably reflecting the needs of the organisms within each lineage. For example, one of the largest gene superfamilies is that of the olfactory receptor (OR) genes.[6] In this superfamily of genes, there has been a net loss of functional OR genes in primates (Figure 3-3), from approximately 660 OR genes in our common ancestor to only 494 functioning OR genes today. Many of the "lost" OR genes are still present in our genome in the form of pseudogenes. It has been suggested that the evolution of trichromatic vision among primates may have lessened the need for an extensive repertoire of OR genes. In contrast, the OR gene superfamily has nearly doubled in membership through time in the lineages leading to mice and rats (see Figure 3-3).

Two important gene families in the metabolism and disposition of drugs have undergone similar changes. The first of these is the cytochrome P450 gene family, which includes some of the most important drug-metabolizing enzymes. Mice have 102 functioning members of this gene family. Humans have approximately half this number; there are 57 sequenced genes and 58 nonfunctioning pseudogenes in the human genome (Table 3-2). Humans compare more "favorably" in the second of these gene families, the ABC transporters, which code for an important family of drug transporters (Table 3-3). The ABC transporter superfam-

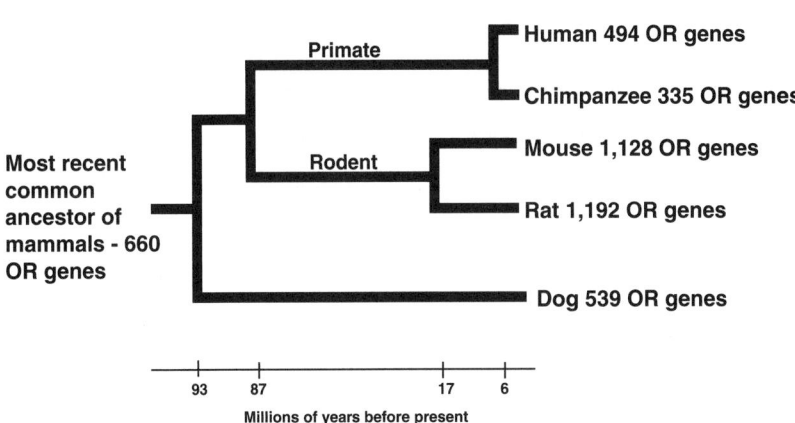

Figure 3-3. An evolutionary tree of olfactory receptor (OR) genes showing the dynamic changes that have occurred among lineages through subgenomic duplication events. The olfactory receptor gene family is among the largest gene superfamilies in mammals, and its members are responsible for the detection of odor molecules. In humans and chimpanzees, many gene members of this family are present in the genome as pseudogenes.[6]

Table 3-2. Members of the Cytochrome P450 Gene Superfamily in Humans*

Humans have 57 sequenced CYP genes and 58 pseudogenes

CYP1A1, **1A2**, 1B1

CYP2A6, 2A7, 2A13, **2B6**, 2C8, **2C9**, 2C18, **2C19**, **2D6**, 2E1, 2F1, 2J2, 2R1, 2S1, 2U1, 2W1

CYP3A4, 3A5, 3A7, 3A43

CYP4A11, 4A22, 4B1, 4F2, 4F3, 4F8, 4F11, 4F12, 4F22, 4V2, 4X1, 4Z1

CYP5A1

CYP7A1, 7B1

CYP8A1, 8B1

CYP11A1, 11B1, 11B2

CYP17

CYP19

CYP20

CYP21A2

CYP24

CYP26A1, 26B1, 26C1

CYP27A1, 27B1, 27C1

CYP39

CYP46

CYP51

Pan troglodytes (chimpanzee)	Probably 57–58
Mus musculus (mouse)	102
Canis familiaris (dog)	54
Bos taurus (cow)	At least 53
Gallus gallus (chicken)	At least 41
Takifugu rubripes (pufferfish)	54
Drosophila melanogaster (fruit fly)	84
Anopheles gambiae (mosquito)	105

*To date, 18 families have been identified in this superfamily. Genes within a family are more similar in sequence than between families and often have more similar functions. Often members of families are located near one another on the chromosome, reflecting their origin as a result of gene duplication events. CYP genes that are important in drug metabolism are shown in boldface type. Inset shows the sizes of this family of genes in other species.

Table 3-3. Members of the Human ABC Transporter Superfamily*

Humans have 48 sequenced ABC genes and 19 pseudogenes

ABCA1, ABCA2, ABCA3, ABCA4, ABCA5, ABCA6, ABCA7, ABCA8, ABCA9, ABCA10, ABCA12, ABCA13

ABCB1 (MDR1), ABCB2, ABCB3, ABCB4, ABCB5, ABCB6, ABCB7, ABCB8, ABCB9, ABCB10, ABCB11

ABCC1 (MRP1), ABCC2 (MRP2), ABCC3 (MRP3), ABCC4, ABCC5, ABCC6, ABCC7 (CFTR), ABCC8,
ABCC9, ABCC10, ABCC11, ABCC12

ABCD1 (ALD), ABCD2, ABCD3, ABCD4

ABCE1

ABCF1, ABCF2, ABCF3

ABCG1, ABCG2 (BCRP), ABCG4, ABCG8

Mus musculus (mouse)	52
Drosophila melanogaster (fruit fly)	56
Yeast	29

*All are ATP-dependent transporters. Many are important drug transporters in the blood–brain barrier, placenta, hepatocytes, renal cells, and endothelial cells. Inset shows the sizes of this family of genes in other species.

ily is one of two large gene families responsible for nearly half of all transporters within cells. ABC transporters function as efflux pumps to move drug molecules out of cells. Members of this family play a crucial role in systemic drug availability, particularly the bioavailability of anticancer drugs. Humans have 48 functional members of this family and 19 pseudogenes. Many of the members of this family have very similar DNA sequences and therefore have significant overlap in the compounds that are transported and the tissues in which they are expressed. This genome complexity is a prominent complicating factor in both drug discovery and pharmacogenetics. With so many very similar genes performing similar, sometimes overlapping functions, it is difficult to identify the specific genes that are essential for a given drug's action. An understanding of the nature and evolution of important gene families will greatly aid in efforts to decipher drug action, identify new drug targets, and reduce drug side effects.

3.2 The Human Genome

Humans have roughly 22,000 to 24,000 genes arrayed on 23 pairs of chromosomes (Table 3-4)—22 pairs of autosomes and the X and Y sex chromosomes. Chromosome size varies from approximately 50 million to 247 million base pairs. Excepting the X chromosome, human chromosomes are numbered by size, chromosome 1 being the largest and chromosome 22 the smallest. The approximately 23,000 human genes are not arrayed on the chromosomes evenly. For example, the 19th chromosome, which represents slightly more than 2% of the total DNA in the cell, has the highest density of genes, containing 1,300 to 1,700 genes (see Table 3-4). In contrast, the larger 18th chromosome, representing 2.7% of the entire genome, has only 300 to 500 genes. The chromosome with the lowest gene density is the Y chromosome—not surprising, because more than half of all human beings do not have a Y chromosome. Interestingly, chimpanzees have 24 pairs of chromosomes, one more pair than humans. The human 2nd chromosome, one of the largest, is thought to be derived from the fusion of two ancestral primate chromosomes.[7]

Aside from approximately 23,000 functioning genes, humans have a nearly equal number (about 20,000) of nonfunctioning pseudogenes. Many of these pseudogenes are located near functioning genes of similar sequence, suggesting that gene duplication, such as unequal crossing over, is the mechanism of their origin. Of the functioning genes, 5% to 10% are RNA genes, including rRNA genes, tRNAs, miRNAs, and snRNAs. The remaining genes are transcribed into mRNAs and translated into polypeptides. The average gene size is approximately 27,000 base pairs on a chromosome (gDNA), including introns and exons (Table 3-5). The average mRNA transcribed from these genes is approximately one-tenth the size of the genomic sequence, at 2,600 base pairs. Intron sizes range

Table 3-4. The Human Chromosome Set*

Chromosome	Size (10⁶ base pairs)	Estimated No. of Genes	Gene Density (genes/10⁶ bp)
1	247	3,000	12.1
2	243	1,300–1,400	5.6
3	200	1,100–1,500	6.3
4	191	1,300–1,600	7.6
5	181	900–1,300	6.1
6	171	1,100–1,600	7.9
7	159	1,150	7.2
8	146	700–1,100	7.2
9	140	800–1,300	8.2
10	135	800–1,200	7.4
11	134	1,500	11.2
12	132	1,200–1,400	9.8
13	114	600–700	5.7
14	106	800–1,300	9.9
15	100	650–1,000	8.3
16	89	850–1,200	11.5
17	79	1,200–1,500	17.1
18	76	300–500	5.3
19	64	1,300–1,700	23.4
20	62	700–800	12.1
21	47	300–400	7.4
22	50	500–800	13.0
X	155	900–1,400	7.4
Y	58	70–200	2.3

* With the exception of the X chromosome, chromosomes are numbered from the largest to the smallest. Gene density on human chromosomes is not uniform but varies by more than 10-fold, from the Y chromosome, a gene "desert," to the 19th chromosome, which has the highest density.

Table 3-5. Statistics for a Few Human Genes*

Gene	Genomic Size (kb)	mRNA Size (bp)	No. of Introns
Insulin	17	400	2
Protein kinase C	11	1,400	7
Albumin	25	2,100	14
Catalase	34	1,600	12
LDL receptor	45	5,500	17
Factor VIII	186	9,000	25
Dystrophin	2,400	17,000	77
Titin (connectin)	283	34,350	362
"Average gene"	27	1,300	0–362

* Note that gene size on the chromosome is in thousands of base pairs (kb). mRNA size is in base pairs (bp). Dystrophin, the gene that, when not functioning correctly, results in muscular dystrophy, is the largest human gene discovered to date, at 2,400 kb (2,400,000 base pairs). The largest mRNA for a human gene is titin.

from 0 to 800,000 base pairs. Although the vast majority of human genes have introns, some genes—including many RNA genes, genes for histones, heat shock proteins, and some receptor genes—lack introns. **Intergenic** regions, that is, the space between neighboring genes on a chromosome, average 75,000 base pairs. Thus the average gene, roughly 2,600 base pairs in size, is fragmented into many introns scattered within a region of DNA 10 times larger than the gene itself and separated from the next gene by a region more than 28 times larger. All of these factors result in a very complex genome. This has made just the discovery of genes, even with the mRNA sequence in hand, very difficult. The functioning portions of human genes are truly very small islands in a very large sea of DNA.

Official human gene names are determined by the Human Genome Organization. Current gene names can be found at http://www.genenames.org/genefamily. html. Gene names are generally short and intended to convey some information as to their function. Generally, members of the same gene superfamily share the same root name, and Arabic numerals are used to distinguish individual members of the family (Table 3-6). In print, gene names are italicized. Pseudogenes may be designated by the suffix "P" after the appropriate gene name. For example, within the cytochrome P450 gene family 2B, there is an identified pseudogene labeled *CYP2B7P*. Given that in humans the rate of "copying errors," or mutation, is approximately 2.5 changes in 10^8 sites per generation per gene, there are many variants segregating among human populations.[8] These variants, or alleles, are designated by the gene name followed by an asterisk and an Arabic numeral; for example, *CYP3A4*2* is one allele or variant (an SNP resulting in the substitution of a "T" for a "C" in some individuals; see Table 3-6) of at least 20 alleles identified to date in this gene (Table 3-7). Allele designation may also include additional information about the specific mutation. Gene duplications resulting in multiple functioning copies of the gene may include the two-letter designation "UM" (see Table 3-6). For example, among humans there are individuals who have 2 to 16 functioning copies of *CYP2D6* and are ultra-rapid metabolizers. All of the variants

Table 3-6. Allelic Nomenclature for a Number of Important Alleles in Pharmacogenetics

Nomenclature	Gene	Protein	Description	Allele Names
ABCB1 3435 C>T (db SNP rs1045642)	Adenosine tri-phosphate–bind-ing cassette pro-tein B1, multiple drug resistance gene (MDR1)	P-glycoprotein (P-gp)	3,435th nucleotide on DNA C is reference nucleotide, T is variant nucleotide Synonymous substitution, silent site substitution	
CYP2C9 144 Arg>Cys (db SNP rs1799853)	Cytochrome P450 subfamily IIC, polypeptide 9	Mephenytoin 4-hydroxylase	144th amino acid, arginine is reference amino acid, cysteine is variant amino acid, nonsyn-onymous substitution	CYP2C9*2 CYP2C9 430 C>T
TPMT*3C (db SNP rs1142345)	Thiopurine S-methyltransferase	S-adenosyl-L-methionine: thiopurine S-methyltrans-ferase	*3 allele, nonsynonymous substitution associated with decreased enzyme activity	TPMT 874 A>G TPMT 240 Tyr>Cys
CYP2D6 1707 del T (db SNP rs5030655)	Cytochrome P450 subfamily IID, polypeptide 6	Debrisoquine/ sparteine hydroxylase	Deletion of a "T" at position 1707, causes a frameshift mutation and a truncated nonfunctional protein	CYP2D6*6
CYP2D6 UM	Cytochrome P450 2D6	Cytochrome P450 subfamily IID, polypeptide 6	Multiple copies of CYP2D6, has been described with different alleles, ultra-rapid metabolizer	
UGT1A1*28 (db SNP rs8175347)			TA repeat in promoter region results in reduced levels of transcription	UGT1A1*28(TA)$_6$ $\to$ (TA)$_7$
VKORC1 -1639G>A (db SNP rs992321)	Vitamin K reduct-ase complex, subunit 1	Vitamin K epox-ide reductase	SNP in promoter region of VKORC1 gene, 1,639 bp upstream of transcriptional start site	

Table 3-7. Polymorphisms in Human CYP3A4 Chromosomal Location 7q22.1*

ALLELE	Nucleotide changes		Effect
	GENE	cDNA	
CYP3A4*1A	None (wildtype)	None	
CYP3A4*1B	-392A>G		
CYP3A4*1C	-444T>G		
CYP3A4*1D	-62C>A		
CYP3A4*1E	-369T>A		
CYP3A4*1F	-747C>G		
CYP3A4*1G	20230G>A		
CYP3A4*1H	20230G>A; 26206C>A		
CYP3A4*1J	6077A>G		
CYP3A4*1K	-655A>G		
CYP3A4*1L	-630A>G		
CYP3A4*1M	-156C>A		
CYP3A4*1N	14200T>G		
CYP3A4*1P	15727G>A		
CYP3A4*1Q	15809T>C		
CYP3A4*1R	16775A>G		
CYP3A4*1S	17815_17816delAT		
CYP3A4*1T	26013T>C		
CYP3A4*2	15713T>C	664T>C	S222P
CYP3A4*3	23171T>C	1334T>C	M445T
CYP3A4*4	13871A>G	352A>G	I118V
CYP3A4*5	15702C>G	653C>G	P218R
CYP3A4*6	17661_176622insA	830_831insA	277Frameshift
CYP3A4*7	6004G>A	167G>A	G56D
CYP3A4*8	13908G>A	389G>A	R130Q
CYP3A4*9	14292G>A	508G>A	V170I
CYP3A4*10	14304G>C	520G>C	D174H
CYP3A4*11	21867C>T	1088C>T	T363M
CYP3A4*12	21896C>T	1117C>T	L373F
CYP3A4*13	22026C>T	1247C>T	P416L
CYP3A4*14	44 T>C	44T>C	L15P
CYP3A4*15A	14269G>A	485G>A	R162Q
CYP3A4*15B	-845_-844insATGGAGTGA; -392A>G; 14269G>A	485G>A	R162Q
CYP3A4*16A	15603C>G	554C>G	T185S
CYP3A4*16B	15603C>G; 20230G>A	554C>G	T185S
CYP3A4*17	15615T>C	566T>C	F189S
CYP3A4*18A	20070T>C	878T>C	L293P
CYP3A4*18B	20070T>C; 20230G>A	878T>C	L293P
CYP3A4*19	23237C>T; 20230G>A	1399C>T	P467S
CYP3A4*20	25889_25890insA	1461_1462insA	488Frameshift

* CYP3A4 is found on the 7th chromosome, long arm (queue), region 2, subregion 2.1. Sequence locations with negative numbers (allele CYP3A4*1B, for example) indicate polymorphisms located upstream of the coding region in the regulatory region of the gene. Importantly, many of the alleles (CYP3A4*6 and CYP3A4*20, for example) may have the same phenotype, an example of allelic heterogeneity.

have been combined under the designation *CYP2D6 UM*. Allele designations may also be followed by a descriptor for the mutation. For example, a fairly common substitution in the drug efflux pump *ABCB1* occurs at the 3,435th nucleotide. Most individuals have a cytosine at this location (wild type), but some have a thymine. In the designation, the wild type, or reference state, is listed first; thus, *ABCB1 3435 C>T*. Often if the substitution results in an amino acid change, the designation notes the amino acid that has changed, for example, *CYP2C9 144 Arg>Cys*; again, the wild-type variant (arginine) is listed first.

3.3 Pharmacogenomics and Pharmacy Practice

A number of important points bear on dynamic genomes and pharmacy practice. First, the size and complexity of gene families that are important in drug metabolism or disposition or are important drug targets (receptors) often complicate the identification of the actual genes and protein products that are involved in a specific drug response. In many cases, two or more genes code for similar enzymes involved in the metabolism of a single drug. In such cases, where there is more than one gene involved in drug response—whether as the drug target or in metabolism or disposition—the likelihood that a mutation in one of the genes will have important clinical effects is lessened, because the loss of one gene can be compensated for by other genes in the same pathway. Such **polygenic** traits, in which multiple genes determine the clinical outcome, greatly complicate attempts to reliably predict patient response on the basis of genetic tests involving only one gene. The appropriate genetics tests would have to assess the allelic status of all the genes in the pathway.

Slightly different from polygenic traits, where multiple genes interact to affect a trait or clinical outcome, are cases in which two or more genes act independently to determine drug response. For example, in a metabolic pathway with two or more steps, mutation in genes at any step that block the pathway may have the same phenotype or clinical presentation (Figure 3-4). This situation, in which mutations in either gene elicit the same phenotype, is referred to as locus heterogeneity. Locus heterogeneity complicates not only the design of appropriate genetic tests but also the discovery of the independent genes involved in a clinical response. For example, clinical studies may detect a polymorphism in enzyme I in Figure 3-4B that explains poor drug response (phenotype C in Figure 3-4B) in a group of patients. Further studies conducted in a different population to collaborate the first may fail to find a significant relationship between polymorphisms in enzyme I and poor drug response. In this group of patients, the poor drug response is due to polymorphisms in enzyme II, which may be unknown to the clinical researchers.

A. Polygenic

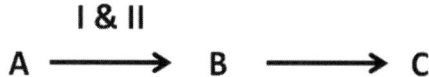

B. Locus heterogeneity

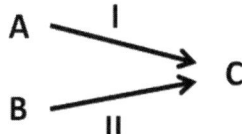

Figure 3-4. Simplified models depicting the difference between traits that are polygenic or display locus heterogeneity. Polygenic traits are the result of two or more genes that interact with one another to produce a given phenotype (C). Traits that display locus heterogeneity are the result of two or more genes that independently give rise to the same phenotype. Note that in genetic mapping studies, locus heterogeneity may confound the results if the investigators do not know that their study population includes individuals whose phenotype is caused independently by two different genes.

Similarly, there are always many alleles for any single gene, many as yet undiscovered. These distinct alleles within one gene may have similar phenotypes (allelic heterogeneity) or different phenotypes (see Table 3-7). For example, *CYP3A4*6* and *CYP3A4*20* are both frameshift mutations resulting from different insertion mutations. Both alleles result in truncation of transcription. Two patients may both be poor responders to a drug and exhibit the same enzyme deficiency as a result of entirely different alleles due to different mutations. A single genetic test may not detect both mutations; one, in fact, may be unknown to science. Locus and allelic heterogeneity may severely limit the universality of genetic tests. This difficulty is compounded by the fact that, in many cases, the frequency of an allele may vary from one ancestral population to the next. Thus genetic tests developed in one group, for example, North Europeans, may not predict the same poor drug response in North Africans.

Finally, in animal studies where the goal is to understand the functioning of a given gene, the complexity of gene families among organisms creates a unique level of difficulty in designing experiments. For example, scientists wishing to understand the role of the cytochrome P450 gene *CYP2C9* in the metabolism of codeine in humans may seek an animal model for experimentation. It is essential for successful experiments to identify the same gene in the experimental animal.

It is necessary to identify the gene in the animal model that metabolizes the same substrates and is expressed in the same tissues. If one wishes to understand human *CYP2C9*, it does little good to study *Cyp1a* in the mouse. The gene most likely to meet these requirements in two different organisms is also likely to share the same evolutionary history, performing the same functions through evolutionary time. These genes, present in both species, are said to be **orthologous**, or orthologs of one another. Identifying the orthologs of important human genes in other model organisms is not straightforward. Identification of an ortholog cannot be based on gene sequence alone, because it is not clear how much sequence difference one should expect between two genes in different species. Ultimately, orthologs must be confirmed by laboratory experiments showing that the two genes serve the same function in both organisms. In addition, the laborious process of identifying the appropriate mouse ortholog of a gene of interest in humans will have to be repeated if further studies are to be done in a different organism or cell line from a different organism. However, knowing something of the evolution of gene families—the evolution of genomes—greatly facilitates the discovery and ultimate validation of genes that are important in human drug response.

Intergenic:
Regions; that is, the space between neighboring genes on a chromosome.

Polygenic:
Traits whose outcome or phenotype are due to the action of two or more genes.

Orthologous:
Genes found in another organism that is homologous to the gene being studied in humans. Homologous genes share a common ancestry and are descended from the same ancestral gene. Members of the same gene family that have different functions in the cell and are not homologous are paralogs.

QUESTIONS

1. The members of the gene family *CYP2D* are clustered on chromosome 22. The cluster contains the functional gene *CYP2D6* and two or three highly homologous pseudogenes. What are pseudogenes, and how do they arise in the genome?

2. True or false: It is possible for a gene to be functional in some human populations and a pseudogene in other populations.

3. The mouse gene *CYP2C65* is believed to be the ortholog of the human *CYP2C19* gene. What is meant by this statement? Suggest ways one might confirm that human *CYP2C19* and mouse *CYP2C65* are orthologs.

4. The mutation *CYP2D6UM* is a multicopy variant present at a frequency of 1% to 5% in Americans and results in ultra-rapid metabolization of codeine. The mutation is always dominant; explain why. If a woman who is an ultra-rapid metabolizer (UM phenotype) is married to a normal male (EM phenotype), what proportion of their children will be UM?

5. The mitochondrial genome is different from the nuclear genome in a number of instances. Describe three of these differences other than size.

References

1. Lander ES, Linton LM, Birren B, et al. Initial sequencing and analysis of the human genome. *Nature.* 2001;409:860–921.
2. Venter JC, Adams MD, Myers EW, et al. The sequence of the human genome. *Science.* 2001;291:1304–51.
3. International Human Genome Sequencing Consortium. Finishing the euchromatic sequence of the human genome. *Nature.* 2004;431:931–45.
4. Taft RJ, Pheasant M, Mattick JS. The relationship between non-protein-coding DNA and eukaryotic complexity. *Bioessays.* 2007;29:288–99.
5. Dehal P, Boore JL. Two rounds of whole genome duplication in the ancestral vertebrate. *PLoS Biol.* 2005;3:e314.
6. Demuth JP, Bie TD, Stajich JE, et al. The evolution of mammalian gene families. *PLoS ONE.* 2006;1:e85.
7. De Grouchy J. Chromosome phylogenies of man, great apes, and Old World monkeys. *Genetica.* 1987;73:37–52.
8. Nachman MW, Crowell SL. Estimate of the mutation rate per nucleotide in humans. *Genetics.* 2000;156:297–304.

Genomic Technologies and Pharmacogenomics

LEARNING OUTCOMES:

At the end of this chapter, you should be able to:

1. Discuss the importance of restriction endonucleases and the Sanger "dideoxy" method of DNA sequencing. How have these discoveries had a significant impact on genomic technologies?

2. Define and discuss the importance of the following terms as they relate to genomic technologies:

   a. Vector
   b. Base coverage
   c. NGS
   d. Chimeric
   e. Transgene

3. Discuss and outline the microarray process and its potential significance in the pharmaceutical and biomedical sciences and clinical practice.

4. Compare and contrast the differences between the genomic methodologies of PCR and real-time PCR and the types of samples that can be analyzed with these technologies.

5. Define and describe the use of silencing RNAs and their potential uses in the pharmaceutical and biomedical sciences and clinical practice.

6. List, compare, and contrast the main genomic and pharmacogenomic databases.

In many fields of science, advances in technology drive the pace of research. This is particularly the case in genomics and pharmacogenomics. Recent technological advances hold great promise for understanding the molecular mechanisms that underlie drug action. This understanding will allow us to identify new drug targets, improve the ones we have, and elucidate the role of genetics in interpatient variability in disease and drug response.

The pace of technological advances has far exceeded scientists' ability to analyze the massive amounts of data now being collected. For example, although our knowledge of the workings of the human genome is still in its very early stages, the ability of the newest DNA sequencers to generate sequences is overwhelming. In 2001, the initial rough draft of the first human genome was released. The sequencing effort took approximately 5 years and cost $2.7 billion. In 2010, the sequences of three individuals were published, the result of a 6-month effort at a cost of approximately $50,000 each. In the intervening 10 years, there was a 10-fold reduction in time and more than a 50,000-fold reduction in cost. The goal of the National Human Genome Research Institute is to sequence a human genome for $1,000.

The field of bioinformatics has greatly expanded to address the deficiencies in data handling, annotation, and statistics of the immense datasets being generated from these technologies. Still, given our extremely rudimentary knowledge of life's workings at the molecular level, the field of pharmacogenomics must be viewed as being in its infancy. It is hoped that continued work will provide the understanding needed to unravel the nature of medical therapeutics and the diseases that trouble humankind. In the meantime, some working knowledge of the technologies involved is necessary in order to assess the significance and limitations of this research in its application to clinical practice.

4.1 First Tools in Molecular Biology

The discovery of the cellular tools necessary to cut or restrict DNA initiated the revolution in molecular biology and gave rise to the field of genomics. Although the skills necessary to isolate relatively pure DNA are fairly straightforward, the difficulty in isolating the same intact, specific region of DNA from a large genome hindered progress in molecular biology for many years following the discovery of the structure and components of DNA. The genome is massive, and the ability to reliably and repeatedly cut DNA at specific locations and recover specific fragments with known sequences at each end proved to be the breakthrough that allowed scientists to begin to partition the genome into sections small enough to study. These restriction enzymes, or endonucleases, are isolated from bacteria and recognize short, **palindromic** sequences of double-stranded DNA and cut the DNA, leaving behind either blunt or staggered ends (Figure 4-1). Restriction enzymes, of which many hundreds are now commercially available, are probably a defense mechanism in bacteria against invading viruses. With the ability to reliably produce DNA fragments with known staggered ends, it quickly became possible to recombine fragments by using the overlapping ends to facilitate the joining or ligation of the DNA fragments (Figure 4-2). Typically, DNA fragments are ligated into specialized plasmids as **vectors** for the DNA fragment to allow for further study, replication, or sequencing (Figure 4-3).

The next major advance in molecular biology came with the discovery of a method to determine the sequence of DNA. The most common technique used was first described by Sanger et al.[1] The Sanger "dideoxy" method for sequencing took advantage of the fact that special nucleotides (dideoxynucleotide triphosphates), when incorporated into a growing DNA strand, stop the continued addition of more nucleotides. Thus a standard sequencing reaction requires:

1. The single-stranded DNA molecule to be sequenced and DNA polymerase to synthesize the new complementary strand.

2. A short DNA primer to initiate the sequencing reaction.

3. The necessary nucleotides for new strand synthesis (deoxyadenosine triphosphate [dATP], deoxycytosine triphosphate [dCTP], deoxy-guanosine triphosphate [dGTP], and deoxytyrosine triphosphate [dTTP]).

4. One each of the four DNA chain terminators (ddATP, ddCTP, ddGTP, and ddTTP).

5. Thermostable DNA polymerase to synthesize a new strand complementary to the target.

Sequencing reactions are typically repeated many times in a thermocycler and yield four reactions, each containing DNA fragments that all begin at the same priming site but end at different sites, depending on the chain terminators. For example, in the reaction tube with the chain terminator ddATP, there are many DNA fragments that all have the same starting sequence, but each of which ends whenever an A occurred in the original sequence.

The DNA fragments in the four reaction tubes are then size fractionated from the smallest to the largest fragments, one base at a time. The size of each fragment is determined by the ending nucleotide of sequence. Thus the DNA sequence

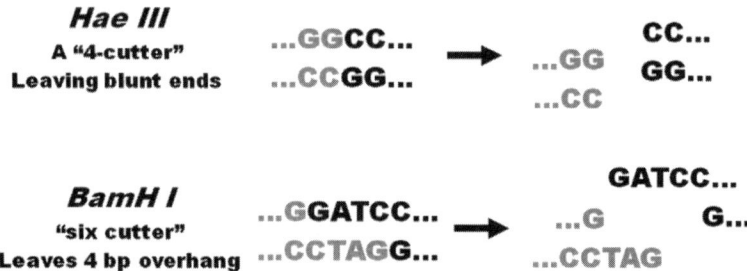

Figure 4-1. Restriction endonucleases, or restriction enzymes, used to cut DNA at specific, sequence-dependent sites. *Hae III*, isolated from *Haemophilus aegypticus*, recognizes a four–base-pair sequence and leaves blunt ends on the two DNA strands. *BamH I*, isolated from *Bacillus amyloliquefaciens*, has a six–base-pair recognition site and leaves staggered, or "sticky," ends. The overhanging ends facilitate the subsequent cloning fragments with complementary ends via a ligation reaction.

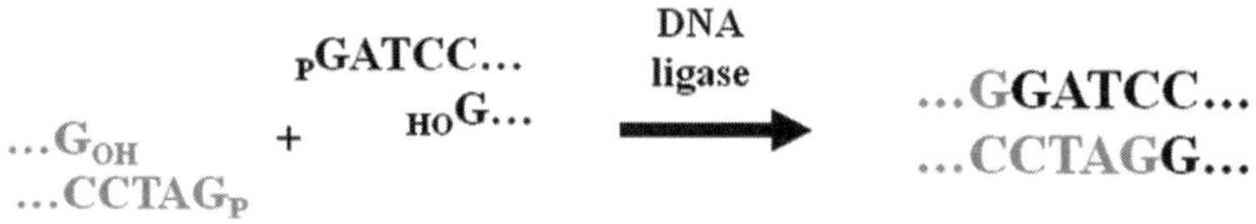

$$\ldots G_{OH} \qquad _pGATCC\ldots \qquad \xrightarrow[\text{ligase}]{\text{DNA}} \qquad \ldots GGATCC\ldots$$
$$\ldots CCTAG_p \qquad + \qquad _{HO}G\ldots \qquad\qquad\qquad \ldots CCTAGG\ldots$$

Figure 4-2. Ligation reaction resulting in the joining of two DNA fragments with complementary overhangs. The enzyme DNA ligase requires that the DNA fragments have the 5' phosphate groups.

can be "read off" beginning with the smallest fragment. Recent next-generation sequencing (NGS) systems employ a number of strategies, but most immobilize the DNA to be sequenced to a solid substrate, allowing for the amplification of the target DNA and the simultaneous sequencing of many thousands to billions of DNA strands. The new emerging NGS technologies will ultimately make the cost and speed of genome sequencing comparable to those of many standard medical tests. The ability to sequence a patient's entire genome or a patient's tumor genome may revolutionize health care. As noted previously, the major limitation to utilizing these technologies is our lack of knowledge about the relationship between genetic data and human health.

Another major advance in molecular biology came with the discovery of a method to amplify a specific DNA fragment by a billion-fold within a heterogeneous mix of DNA. Polymerase chain reaction (PCR) employs short, specific primers that anneal to the opposing strands of denatured, double-stranded DNA and a thermostable DNA polymerase to synthesize new complementary strands. After the first round of synthesis, the DNA strands are denatured by heating to greater than

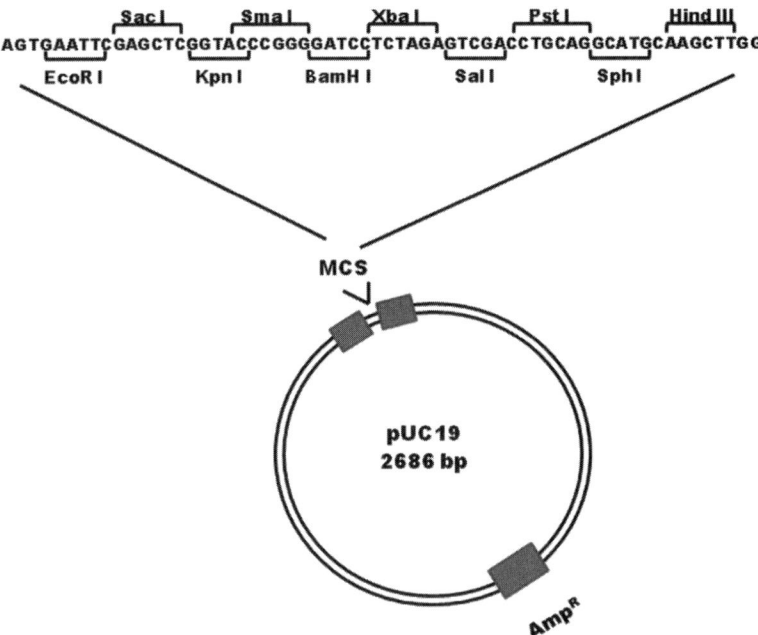

Figure 4-3. Plasmid cloning vector with the multiple cloning site expanded to show the restriction sites available to allow cloning of a DNA fragment with the appropriate complementary ends. The plasmid also includes a gene conferring ampillicin resistance to the bacterial cell that incorporates the plasmid. This allows for the selection of bacterial colonies that carry the plasmid when cells that have been transformed with the plasmid are grown in media with ampillicin.

92°C (198°F), followed by another round of primer annealing and DNA synthesis. The cycle is repeated 20 to 40 times, and each time the amount of target DNA is roughly doubled (Figure 4-4). PCR machines are nothing more than heat blocks that can be programmed to repeat a series of thermocycles (Figure 4-5). The primers are designed to be complementary and specific to the region of DNA to be amplified. Lengths of 20 base pairs or more are enough to ensure that the sequence occurs only once in even the largest genomes. In this way, it is possible to isolate and amplify a specific piece of DNA from a complex genome in a few hours or less.

4.2 Genomic Technologies

4.2.1 Massively Parallel DNA Sequencing
Sanger sequencing has dominated sequencing efforts for the past 20 years. Recently, new sequencing technologies have been developed that greatly increase the amount of DNA sequence that can be generated in a single experi-

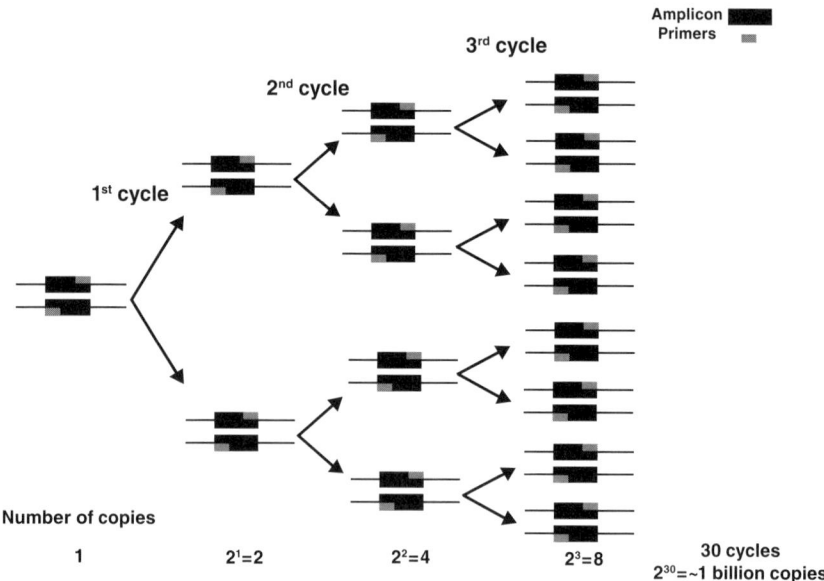

Figure 4-4. Amplification of DNA using PCR. Note that only the DNA flanked by the primers is amplified with each cycle. Thus, with the appropriate primers, specific DNA sequences can be isolated from a heterogeneous mixture of DNA.

Initial denaturation	90° – 95° C	1 – 3 min
Denature	90° – 95° C	0.5 – 1 min
Primer annealing	45° – 65° C	0.5 – 1 min
Primer extension	70° – 75° C	0.5 – 2 min
Final extension	70° – 75° C	5 – 10 min
Stop	4° C	hold

25 – 40 Cycles

Figure 4-5. The PCR thermocycle. The denaturing step "breaks" the hydrogen bonds between the two DNA strands, thus allowing the primers to anneal to their complementary target sequences during the primer annealing step. The attached primers then serve as the substrate for a thermostable DNA polymerase, which copies the DNA strand by using the original DNA as a template. These steps, from denaturation to primer extension, are repeated up to 40 times. The amplification cycles are often followed by a final extension step to allow the DNA polymerase to "clean up" any staggered ends.

ment. The increase in sequence data generation is accomplished by massively parallel sequencing reactions that enable up to billions of sequencing reactions to occur simultaneously in a single run.[2,3] Although the technologies vary in elements of sample preparation and sequence detection, these NGS technologies will change many aspects of basic and clinical research.[4–6] At present, all NGS technologies are limited by the short sequence reads obtained, ranging from 25 to 350 base pairs. Such short read lengths—compared to the 500-to-750–base pair read lengths from Sanger sequencing—necessitate significant computational horsepower to perform the assembly of longer stretches of DNA. Also generally required is prior sequence information to allow for alignment. Sequences can be aligned de novo, but this requires much more **base coverage**, that is, greater than 30-fold.

Aside from the addition of new genomes from species as yet unsequenced, the NGS efforts are directed at sequencing individual or "personal" genomes from volunteers, with the goal of cataloging single-nucleotide polymorphisms (SNPs) and insertion/deletion variants in human populations. The ultimate identification and delineation of these variants in human populations are critical to understanding the underlying genetic causes of human disease and drug response. Realization of this goal will require not only immense sequence data from a large number of individuals but also a far better delineation of the underlying molecular causes of human disease and drug response. Because variation in human drug response is in general a far simpler phenotype than even the simplest of diseases, there is justified hope that progress in pharmacogenetics will be seen in the near future. As an example, consider the multitude of known polymorphisms of *CYP3A4* (see Table 3-7). With the advent of NGS systems, the ability to determine the allelic status for all the known variants of a gene is becoming more likely.

Another important arena where NGS will have an immediate impact is in studies of gene expression.[7] Typically, such studies involve collecting messenger RNAs (mRNAs) from tissue samples or cells of interest and then determining which mRNAs have increased or decreased in abundance in relation to control samples.

Palindromic:
A sequence on a DNA strand that has the same sequence on the forward strand as on the reverse strand.

Vector:
Vehicles used to move DNA fragments from one entity to another, for example, from one cell to a target cell. Depending upon the need, vectors can be plasmids, viruses, or artificial chromosomes.

Base coverage:
The average number of reads per nucleotide. For example, eightfold base coverage means that, for a given DNA sequence, each base has been independently sequenced eight times. Base coverage, and therefore sequencing effort, will decrease as the quality of sequence data and read length increase.

Massively parallel sequencing will allow for the sequencing of the entire mRNA pool (via cDNAs) within a sample. Potentially, all the mRNAs in a sample can be quantified, including those that are as yet unknown. This is a major improvement over present-day DNA microarrays (see next section). In addition, the short read lengths of current NGS technologies are not a hindrance, because sequence reads of 50 base pairs or more are adequate to identify most mRNAs.

4.2.2 DNA Microarrays

In general terms, DNA microarrays, or chips, allow for the assessment of the DNA segments that are present in a given sample. Densities of DNA "targets" on a single array or chip can range from hundreds to well over 100,000 short DNA sequences. These arrays take advantage of the fact that the two complementary sequences of DNA will bind, or hybridize, to one another. Thus if one-half of the DNA sequence is covalently bound to an array and its complement is present in a sample to be assayed, the two will bind to one another. This allows one to assess the DNA segments that are present in any sample, given a DNA chip with the appropriate DNA targets. Typically, the sample DNA sequences to be assayed are labeled using an isotope (^{32}P- or ^{33}P-labeled dCTP), fluorescent dyes (Cy3 or Cy5), or chemiluminescence (often, biotin-labeled dUTPs). These array experiments yield image files containing intensity values (minus some assessment of background) resulting from the hybridization of the labeled DNAs to the targets on the array.

The information garnered from a microarray experiment depends on the nature of the targets printed on the arrays. As noted previously, the arrays consist of short segments of DNA that are typically 20 to 70 base pairs in length. In most pharmaceutical research, the arrays are used to address three basic questions:

1. **Gene expression**—Gene arrays are used in studies to identify and characterize patterns of gene expression, which can provide insight into the underlying biochemical and molecular mechanisms of drug response or enable characterization or identification of a tissue or tumor sample. Assessment of the many-fold transcriptional events that are involved in drug response provides the initial building blocks for more mechanistically based studies of the molecular and physiological basis of the effects and disposition of therapeutic agents. Assessments of gene expression can be genome-wide, encompassing all the known genes in the organism. In many cases, the arrays are designed to be pathway specific; for example, arrays can be used to assess the activity of the many hundreds of drug-metabolizing enzymes in a liver sample. In either case, mRNA is isolated from the tissue of interest and generally reverse transcribed into labeled cDNAs. In some cases, the labeled cRNAs are used.

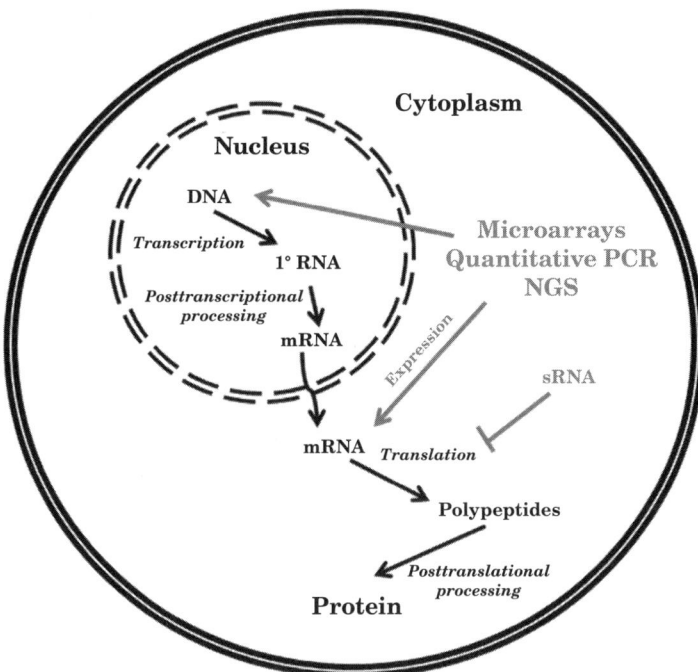

Figure 4-6. Molecular pathway from genomic DNA to final protein products. The major genomic technologies are shown at their site of study. DNA testing or genotyping is carried out with genomic DNA. Gene expression studies are conducted by assessing the mRNA pool with microarrays, quantitative PCR, or NGS technologies. Finally, the functional role of individual proteins can be examined by using silencing RNA technologies that inhibit translation.

These experiments have two major limitations. First, the component being measured or assessed is the mRNA pool in the sample, not the actual functioning proteins (enzymes, transporters, receptors, etc.). The underlying hope in these experiments is that the makeup of the mRNA pool is a valid estimate of the functional protein pool in the cell (Figure 4-6). The relationship between assessed mRNA levels as identified in microarray experiments and the concentrations of the respective proteins may be complex or, more often, unknown.[8–10] However, gene expression experiments provide a mechanism to potentially identify all the molecular players in the cell's response to a therapeutic agent. In many cases, the genes identified in the course of a microarray experiment were not suspected to play a role in the response. A second limitation is that microarray experiments provide only a short, "snapshot" view of the mRNA pool at the instant when the sample was collected. Many samples taken over time are ultimately necessary to gain a complete picture of the cellular response to a given therapeutic agent.

Microarrays have become important tools in assessing gene expression responses in chemotherapy,[11,12] development of drug resistance,[13,14] pharmacodynamics,[15,16] and transcriptional profiling to monitor patient outcomes and drug efficacy.[17–19] Finally, gene expression arrays can be used to provide a "gene expression fingerprint" of a given sample. Here the goal is to use the specific expression pattern to identify a specific tumor type or other disease state in a patient.

2. **Genetic tests**—Microarrays may be composed of a battery of potential SNPs or length variants that are known to be segregating in human populations.[20] From a single sample of genomic DNA, many hundreds of polymorphisms can be assessed across many hundreds of genes.[21] This application is becoming increasingly important as the number of known polymorphisms increases for each gene, for example, *CYP3A4* (see Table 3-7). For these arrays, genomic DNA is isolated and labeled rather than mRNAs, as is the case for gene expression arrays. Given the high density of DNA targets on a typical chip (100,000 or more), it is possible to conduct a genetic test for a battery of known polymorphisms across many genes.

3. **Comparative genomic hybridization**—Microarrays may be composed of human chromosomal segments to allow assessment of the nature of the genomic reorganization present in a tumor sample. This information can be used to identify a tumor type and may provide an understanding of the drastic changes in gene regulation and expression that are observed in many tumors.

Independent of the type of array, the statistical methods applied to microarray analyses can be divided into two broad classifications reflecting two desired outcomes. The first group of analyses concerns the identification of a subset of DNA sequences—cDNAs, SNPs, or chromosomal segments—that exhibit significantly altered quantities due to the experimental conditions. The second broad group of analyses, limited largely to gene expression studies, are often referred to as "data mining" and use intensive computational methods (clustering algorithms, self-organizing maps, principal-component analysis) to identify patterns of gene expression. Such data mining can have one of two goals. In many cases, the goal is to classify, or cluster, genes into groups showing similar patterns of expression. Because these techniques do not provide any additional information to the computational algorithm other than the relative expression levels, they are referred to as "unsupervised approaches." Examples include cluster analysis,[22] k-means clustering,[23] and self-organizing maps.[24] In contrast to these clustering methods, researchers often seek to group samples into functional classes (for example, diseased versus normal tissues) on the basis of gene expression patterns. The ultimate goal is to identify a subset of genes that will allow the researcher to make predictions or assignments of unknown samples on the basis of gene expression patterns. Because the algorithms are provided with both relative-quantity data as well as additional data in terms of classification groups, such methods are referred to as "supervised approaches."[25]

4.2.3 Quantitative Real-Time PCR

As noted previously, PCR has been a standard tool in the detection of specific cDNAs in a sample. Recent technologies for assessing specific quantities of specific targets are based on the ability to monitor the PCR in real time. Through the use of DNA dyes (e.g., SYBR Green I) or fluorogenic probes, the amount of DNA accumulating with each cycle of the PCR amplification process can be quantified. Monitoring DNA quantities in real time allows researchers to estimate the initial starting concentration of a given target based on the amplification profiles. From tissue samples as small as a few hundred cells, it is possible to estimate the number of mRNAs present for any given gene of interest.

Real-time PCR has a number of advantages over microarrays for assessing gene expression. First, the amount of tissue necessary for real-time PCR is an order of magnitude less than that required to carry out a microarray study. It is therefore possible to assess the cellular response to a drug rather than the response of the whole tissue. Second, real-time PCR is faster, less expensive, and more quantitative than microarrays. The major limitation of real-time PCR is the limit to how many genes can be assayed at one time. Currently available PCR arrays can assay up to hundreds of genes in one experiment. This is significantly less than the thousands of genes that can be assayed on a standard microarray.

Real-time quantitative PCR has been used to examine the kinetics of gene expression in response to specific drugs and the role of drug transporters and/or metabolizing enzymes in drug distribution. The technology is also rapidly becoming an important means of genotyping for SNPs, using allele-specific primers. Finally, one of this technology's primary roles is in confirming and extending the findings of microarray studies. Because PCR requires only minute samples, it is possible to assess gene expression from specific tissues or cells. The reduced costs of PCR assays relative to microarrays and NGS also make it possible to monitor gene expression across many more time points under different dosage regimens. However, the same caveat that applies to other gene expression data based on measurements of the mRNA pool also applies to quantitative PCR data: The mRNA pool is dynamic, and a single measurement in time may not be informative about the proteins present in a cell (Figure 4-6).

4.2.4 Silencing RNAs

One of the most recent technologies to become available to researchers also has immense therapeutic potential and is based on **RNA interference.** This phenomenon was discovered in the course of experiments undertaken to produce petunias with more intense purple flowers. Attempts were made to overexpress one of the genes (chalcone synthase [*CHS*]) responsible for the purple color by introducing an additional copy of *CHS*. This resulted in the loss of color in the **chimeric** plants. Further work showed that the introduction of the **transgene** into the petunia cells resulted in suppression of both the transgene and the endogenous gene.[26] The phenomenon has been called "cosuppression," and because suppression occurred at the level of the gene transcripts, it was deduced that the "silencing" had occurred posttranscriptionally. In an excellent example of serendipity in science, what had been discovered was a naturally occurring cellular mechanism of genetic control that silences the expression of a gene with small double-stranded RNAs (dsRNAs) to specifically degrade complementary mRNAs.[27] These naturally occurring dsRNAs are abundant in nearly all genomes and are thought to regulate many hundreds of genes in humans, including many that are important in drug metabolism and drug transport. The endogenous silencing RNAs are called microRNAs (miRNAs) and are thought to play a role in cellular defense against viral infection.

RNA interference:
A cellular mechanism of gene silencing mediated by short double-stranded RNA (dsRNA) molecules that are taken up by the RNA-induced silencing complex with the cytoplasm of the cell and together inhibit the translation of specific mRNAs based upon the sequence of the short dsRNA.

Chimeric:
An organism that contains genetic material from a different species.

Transgene:
A gene or genetic material that has been transferred from one species into another.

Synthetically produced small interfering RNAs (siRNAs) can be designed for nearly any human gene and are processed within the cell by the same cellular machinery used for miRNAs, resulting in the temporary inhibition of the translation of the complementary mRNA. siRNAs have become invaluable as a research tool to silence individual genes in order to study their function. They have proven to be highly specific, silencing only the mRNA for which they have been designed, and to work in a catalytic manner. For example, it is now possible, in order to assess the role of a specific cellular transporter in the movement of a new compound across a cell membrane, to temporally inhibit the translation of the transporter in the given cell line. siRNAs for virtually all human, rat, and mouse genes are now commercially available. Aside from their immediate impact in research and drug development, siRNA technologies have great potential as therapeutic agents. This is particularly the case for the treatment of diseases caused by the inappropriate expression or overexpression of specific genes. In such cases, the ability to inhibit the expression of the causative gene could provide immediate and direct benefit. For example, the inhibition of the genes responsible for cellular proliferation in tumors has great potential in the treatment of cancers.[28] At this time, the major barrier to the development of therapeutic siRNAs is delivery and targeting. There are significant obstacles to be overcome in the understanding of the pharmacokinetics and biodistribution of siRNA carriers (generally, lipids and bioconjugates) and in the ability to target specific tissues.[29,30]

4.3 Genomic and Pharmacogenomic Databases

The best source for timely information concerning genetic polymorphisms and drug response are the databases accessible via the Internet. As might be expected given the sheer size of the Internet, there are a great many websites with pharmacogenomic or pharmacogenetic information. The two databases with the most comprehensive, up-to-date, and trustworthy pharmacogenomic and pharmacogenetic data are the National Center for Biotechnology Information (NCBI) and the Pharmacogenomics Knowledge Base (PharmGKB). Both databases include both primary data and literature and annotated databases. Both also have links to helpful online tutorials and glossaries for navigating through the immense amount of content, and both are free to the public.

NCBI (www.ncbi.nlm.nih.gov) is part of the National Institutes of Health and was created in 1988 as the ultimate repository for all molecular biology information.[31] Largely known as the home for GenBank, which receives DNA sequence data from around the world (in collaboration with the European Molecular Biology Laboratory and the DNA Database of Japan), NCBI is also responsible for the collection and dissemination of many other forms of molecular and genetic data. NCBI has a highly integrated search and retrieval system called "Entrez," which links all the databases.

Entrez allows seamless movement from one database to another within NCBI. With a click of the mouse, users can go from an mRNA sequence (GenBank), to its location within the genome (Genome database), to information about the organism (Taxonomy database), to its three-dimensional structure (3D Structures database), to the literature (PubMed). The database most relevant to human disease, drugs, and genetics is Online Mendelian Inheritance in Man (OMIM). The OMIM database is the repository for all data concerning human genes and genetic diseases. OMIM has textual information and references with links to Medline and all the many additional related resources at NCBI, including sequence data, genetic maps, and SNP data. OMIM is searchable by disease name, gene name, or drug. As of 2010, the database had records for more than 13,161 human genes. OMIM is designed for use primarily by physicians and other health professionals, though some knowledge of genetic terminology and concepts is essential.

PharmGKB (www.pharmgkb.org) is an integrated database providing clinical, pharmacokinetic, pharmacodynamic, genotypic, and molecular function data for human genetic polymorphisms and drugs.[32,33] PharmGKB is free and open to the public and is the best source for up-to-date information concerning genes and drug response. The stated aim of this database, which is developed and maintained by Stanford University, is "to aid researchers in understanding how genetic variation among individuals contributes to differences in reactions to drugs." The data within PharmGKB are classified into five categories of pharmacogenetic knowledge:

1. **Clinical Outcome**—The role of genetic variation in altering clinical endpoints that aid in determining medical practice or policy.

2. **Pharmacodynamics and Drug Response**—The role of genetic variation in differences in the biological or physiological response to drugs. In some situations, these differences may be treated as surrogates for clinical responses.

3. **Pharmacokinetics**—The role of genetic variation in the absorption, distribution, metabolism, or elimination of a drug.

4. **Molecular and Cellular Functional Assays**—The role of genetic variation in laboratory assays for molecular or cellular responses to drugs.

5. **Genotype**—Data on the type and location of genetic polymorphisms relevant to drug action.

The PharmGKB database was established to provide researchers and health professionals with easy access to current data on drugs and genetic variability at a number of levels. Like the information at NCBI, these data represent the most current information. To make the most of the information, some understanding of genetic principles is necessary.

The importance of these databases cannot be overemphasized. The speed at which the field of pharmacogenomics is growing and the amount of data that are being generated make NCBI and PharmGKB indispensible resources for up-to-date pharmacogenomic information. All health care professionals should be familiar with the use of these repositories of genetic information.

QUESTIONS

1. What are restriction endonucleases? Why are they so useful for molecular genetic studies?

2. Briefly describe DNA microarrays. Describe one potential use in the pharmaceutical sciences or clinical sciences.

3. Briefly describe some of the limitations of NGS technologies.

4. Describe the difference between PCR and real-time PCR.

5. In conducting a gene expression study (either DNA microarrays or real-time PCR) versus a genotyping experiment (SNP analysis), there are two important differences in the type of tissue samples to be collected for the analysis. Describe these differences.

6. Describe the resources available at PharmGKB.

References

1. Sanger F, Nicklen S, Coulson AR. DNA sequencing with chain-terminating inhibitors. *Proc Natl Acad Sci USA.* 1977;74:5463–7.
2. Ansorge WJ. Next-generation DNA sequencing techniques. *N Biotechnol.* 2009;25:195–203.
3. Metzker ML. Sequencing technologies—the next generation. *Nat Rev Genet.* 2010;11:31–46.
4. Tucker T, Marra M, Friedman JM. Massively parallel sequencing: the next big thing in genetic medicine. *Am J Hum Genet.* 2009;85:142–54.
5. Voelkerding KV, Dames SA, Durtschi JD. Next-generation sequencing: from basic research to diagnostics. *Clin Chem.* 2009;55:641–58.
6. Aparicio SA, Huntsman DG. Does massively parallel DNA resequencing signify the end of histopathology as we know it? *J Pathol.* 2010;220:307–15.
7. Wang Z, Gerstein M, Snyder M. RNA-Seq: a revolutionary tool for transcriptomics. *Nat Rev Genet.* 2009;10:57–63.
8. Gygi SP, Rochon Y, Franza BR, et al. Correlation between protein and mRNA abundance in yeast. *Mol Cell Biol.* 1999;19:1720–30.
9. Chen G, Gharib TG, Huang C-C, et al. Discordant protein and mRNA expression in lung adenocarcinomas. *Mol Cell Proteomics.* 2002;1:304–13.
10. Kuo WP, Jenssen TK, Butte AJ, et al. Analysis of matched mRNA measurements from two different microarray technologies. *Bioinformatics.* 2002;18:405–12.
11. Nawrocki S, Skacel T, Brodowicz T. From microarrays to new therapeutic approaches in bladder cancer. *Pharmacogenomics.* 2003;4:179–89.
12. Wajapeyee N, Somasundaram K. Pharmacogenomics in breast cancer: current trends and future directions. *Curr Opin Mol Ther.* 2004;6:296–301.
13. Wang EQ, Lee WI, Brazeau D, et al. cDNA microarray analysis of vascular gene expression after nitric oxide donor infusions in rats: implications for nitrate tolerance mechanisms. *AAPS PharmSci.* 2002;4:E10.

14. Akiyama S. The new mechanisms and reversal of drug resistance. *Gan To Kagaku Ryoho.* 2003;30:1–8.

15. Almon RR, DuBois DC, Brandenburg EH, et al. Pharmacodynamics and pharmacogenomics of diverse receptor-mediated effects of methylprednisolone in rats using microarray analysis. *J Pharmacokinet Pharmacodyn.* 2002;29:103–29.

16. Burczynski ME, Oestreicher JL, Cahilly MJ, et al. Clinical pharmacogenomics and transcriptional profiling in early phase oncology clinical trials. *Curr Mol Med.* 2005;5:83–102.

17. Gunther EC, Stone DJ, Gerwien RW, et al. Prediction of clinical drug efficacy by classification of drug-induced genomic expression profiles in vitro. *Proc Natl Acad Sci USA.* 2003;100:9608–13.

18. Simon R. Using DNA microarrays for diagnostic and prognostic prediction. *Expert Rev Mol Diagn.* 2003;3:587–95.

19. Liljedahl U, Karlsson J, Melhus H, et al. A microarray minisequencing system for pharmacogenetic profiling of antihypertensive drug response. *Pharmacogenetics.* 2003;13:7–17.

20. Ji M, Hou P, Li S, et al. Microarray-based method for genotyping of functional single nucleotide polymorphisms using dual-color fluorescence hybridization. *Mutat Res.* 2004;548:97–105.

21. Chicurel ME, Dalma-Weiszhausz DD. Microarrays in pharmacogenomics—advances and future promise. *Pharmacogenomics.* 2002;3:589–601.

22. Eisen MB, Spellman PT, Brown PO, et al. Cluster analysis and display of genome-wide expression patterns. *Proc Natl Acad Sci USA.* 1998;95:14863–8.

23. Ringner M, Peterson C, Khan J. Analyzing array data using supervised methods. *Pharmacogenomics.* 2002;3:403–15.

24. Toronen P, Kolehmainen M, Wong G, et al. Analysis of gene expression data using self-organizing maps. *FEBS Lett.* 1999;451:142–6.

25. Li J, Pankratz M, Johnson JA. Differential gene expression patterns revealed by oligonucleotide versus long cDNA arrays. *Toxicol Sci.* 2002;69:383–90.

26. Napoli C, Lemieux C, Jorgensen R. Introduction of a chimeric chalcone synthase gene into petunia results in reversible co-suppression of homologous genes in trans. *Plant Cell.* 1990;2:279–89.

27. Fire A, Xu S, Montgomery MK, et al. Potent and specific genetic interference by double-stranded RNA in *Caenorhabditis elegans. Nature.* 1998;391:806–11.

28. Takeshita F, Ochiya T. Therapeutic potential of RNA interference against cancer. *Cancer Sci.* 2006;97:689–96.

29. Wang J, Lu Z, Wientjes MG, et al. Delivery of siRNA therapeutics: barriers and carriers. *AAPS J.* In press.

30. Singh SK, Hajeri PB. siRNAs: their potential as therapeutic agents, II. Methods of delivery. *Drug Discov Today.* 2009;14:859–65.

31. Sayers EW, Barrett T, Benson DA, et al. Database resources of the National Center for Biotechnology Information. *Nucleic Acids Res.* 2010;38:D5–16.

32. Altman RB, Flockhart DA, Sherry ST, et al. Indexing pharmacogenetic knowledge on the World Wide Web. *Pharmacogenetics.* 2003;13:3–5.

33. Klein TE, Chang JT, Cho MK, et al. Integrating genotype and phenotype information: an overview of the PharmGKB project. *Pharmacogenomics J.* 2001;1:167–70.

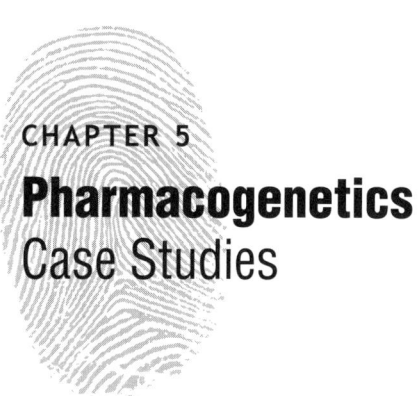

CHAPTER 5

Pharmacogenetics
Case Studies

LEARNING OUTCOMES:

At the end of this chapter, you should be able to:

1. Outline and discuss the importance of environmental responses and their potential impact on pharmacogenetic studies of drug response and disposition.

2. Discuss two critical properties that must be present if pharmaco-genetics is to play a key role in optimizing patient care for a specific therapeutic agent.

3. Predict whether pharmacogenetics may play a key role in optimizing the desired response with a specific drug based on knowledge of the mechanism of action, metabolic or transport properties, and types of gene variants or haplotypes involved in the pharmacological effect and disposition.

It can be argued that the genomics revolution, which has pervaded all of the biological and biomedical sciences, has exhibited its most impressive advances in the pharmaceutical and clinical sciences. As noted in Chapter 4, genomic technological advances are having a tremendous impact from bench to bedside; for example:

■   Next-generation sequencing technologies will soon make it possible to conduct patient genotyping on the scale necessary to evaluate all or most of the known polymorphisms implicated for a given drug.

■   Next-generation sequencing technologies will allow for the quantitative assessment of gene expression for the entire **transcriptome**. The ability to "see" how genes respond to a given therapeutic intervention will allow for the design of safer, more efficacious drugs and therapies. In addition, the molecular understanding of disease to be gained from such research will identify new drug targets. Significantly, these new drugs, directed against specific subtypes of disease, will often require genetic testing in the clinic.

> **Transcriptome:** The set of all RNAs in a cell. These include mRNAs, tRNAs, and rRNAs, though generally mRNAs are the species of interest because mRNAs give rise to proteins.

- Quantitative polymerase chain reaction will allow for quick, inexpensive genotyping, or assessment of gene expression in clinical settings. Speed and ease of use are essential if these technologies are to become standards of care in therapeutic management.
- Silencing RNA technologies that silence specific genes are becoming commonplace research tools to assess the specific role of a given gene (e.g., for drug transporters, metabolizing enzymes) in drug action and disposition. The ability to silence specific genes also has great potential for the development of therapeutic agents.

These genomic tools have contributed to an emerging paradigm shift in pharmacology, pharmaceutics, drug development, and pharmacotherapeutics. The molecular tools that are now readily available to laboratories have hastened the shift from a drug development process that is centered largely on chemistry to one based on our growing biological knowledge of the biochemical and molecular effects of compounds.[1] With these tools, it is now becoming possible to assess the nature of drug action, toxicity, and tolerance at the cellular level. Similarly, our understanding of human disease is being refined, leading to more precise therapeutic interventions based on more systematic understandings of disease states.

These genomic-driven advances in the sciences that underlie pharmacy and medicine will revolutionize health care. Much has been written about the age of personalized medicine, in which each patient's genetic makeup will determine an individual-specific course of drug treatment designed to be the most efficacious and safe. Less well publicized, though likely to have a greater impact, is the use of genomic tools to "individualize disease" or to define common diseases into more precise subtypes based upon specific underlying causative processes. New therapeutic agents can then be developed against the specific targets responsible for the disease subtype. These new therapies are likely to be more efficacious with fewer side effects.

Pharmacogenetics, which has been an active area of research for over 50 years,[2] seeks to provide patient-effective therapies that have minimal adverse drug reactions and are based on the individual's genetic makeup (i.e., genotype).

This goal is based on the concept that it is possible to identify one or more genes that determine drug metabolism and/or transport (drug metabolism or transport pharmacogenetics) or are the direct targets of drug action (drug target pharmacogenetics) to assist clinicians in selecting the appropriate drug for an individual patient. The literature in this area is extensive and includes several new journals that detail studies showing gene–drug relationships and the importance of including a patient's genetic makeup in guiding therapeutic decisions.[3,4] However, a significant portion of this literature consists of contradictory reports or analyses that call into question the utility of genetic testing in guiding therapeutic management. For example, one of the most studied genes in pharmacogenetics, *MDR1* (*ABCB1*)—the gene that codes for the ubiquitous drug efflux pump, P-glycoprotein—has so far defied any straightforward consensus concerning the importance of genetic polymorphisms in drug disposition and response.[5] Regarding the pharmacogenetics of cytochrome P450 genes (*CYP*), Nebert and Vesell[6] have cautioned that even recent, much lowered estimates of reductions in adverse drug reactions of 10% to 20% by extensive genotyping of *CYP* polymorphisms may be overly optimistic. This steady background of contrary data suggests caution in assessing the importance of pharmacogenetics in designing and optimizing drug therapy regimens.[7–10]

That the extensive pharmacogenetic literature is often conflicting does not mean that there is little to be gained from pharmacogenetic studies and testing. Rather, it suggests a need among practicing health care providers for a more realistic understanding of the role of environmental factors, multiple genes with multiple variants, and human population genetic structure in predicting individual drug efficacy and toxicity. This chapter presents two case studies of pharmacogenetic success that highlight the confounding issues associated with genes that play a role in drug response and efficacy. The genomic science covered in the preceding chapters is invoked here to allow for a better understanding of drugs and human response.

5.1 Environmental Factors and Drug Response

Before examining the role of genomic complexities in our understanding of pharmacogenetics and drug response, it is essential to consider the influence of environmental factors in patient variations in drug response. Many environmental factors determine how a given patient responds to a drug. These factors include the patient's age, lifestyle, and diet, as well as concurrent drug therapies (see Figure 1-1). These nongenetic factors often play a large role—generally a greater role than genetics—in a patient's response to a given drug.[10] Identifying the environmental factors involved in drug response is absolutely essential before the role of genetics can be accurately assessed. For example, many statins

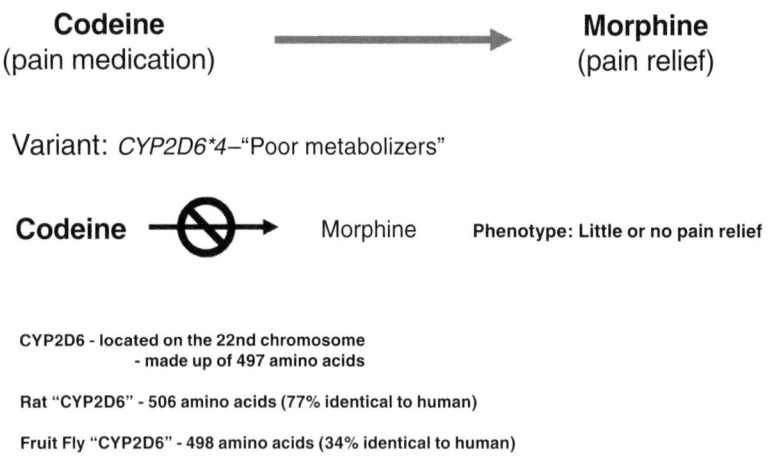

Figure 5-1. Metabolic pathway for the conversion of codeine to morphine in the liver. The cytochrome P450 enzyme responsible for this conversion (*CYP2D6*) is polymorphic among humans. One common variant in Caucasians, *CYP2D6*4*, results in reduced levels of enzyme, particularly in homozygous individuals. As the inset shows, *CYP2D6* is a very old gene that is found in most eukaryotes. Studies of the homologous gene in rats have been important in our understanding of this gene's biochemical function.

(3-hydroxy-3-methylglutaryl–coenzyme A reductase inhibitors) that are taken to lower plasma cholesterol levels are known inhibitors of specific cytochrome P450 enzymes.[11] Specifically, the gene *CYP2D6* codes for the enzyme that is largely responsible for the conversion of codeine to morphine in the liver (Figure 5-1). A polymorphism for this gene, *CYP2D6*4*, is quite common in Caucasians and results in a nonfunctional phenotype.[12] (See Pharmacogenomics Knowledge Base [www.pharmgkb.org] for the latest information on other variants and their frequencies in human populations.) Individuals who are homozygous for this variant derive little or no pain relief upon taking codeine. Thus there is a clear relationship of genotype (*CYP2D6*4*) to phenotype (little or no analgesic response). However, the same phenotype (little or no analgesic response) may also be seen in patients who are taking statins for treating high cholesterol. As can be seen in the following case studies, environmental factors may often confound the relationship between genetics and a patient's given drug response.

5.2 Pharmacogenetic Case Studies

The two pharmacogenetic "success stories" described here are in large part successful because the compounds involved possess two critical properties. The first property is one inherent in the drug itself. Most drugs for which pharmacogenetics is clinically relevant have a narrow therapeutic window. By definition, the safe therapeutic dosage (and blood concentration) range for these drugs is small, and thus slight genetically based differences in drug metabolism, absorption, distribution, or clearance may result in adverse effects. Drugs that show no adverse effects across a wide concentration range are unlikely to have pharmacogenetic issues, because most genetic differences in metabolism, absorption, or distribution will be within the range of the normally administered dose.

The second property is inherent in the biological pathways involved in the compound's activation or inactivation. Drugs that have "pharmacogenetic" indications have at least one critical step in the drug response pathway that is controlled principally by a single gene—the phenotype is said to be monogenic (see Chapter 2). Inactivation of this single limiting gene may then have phenotypic consequences. In contrast, drugs that are metabolized by many different genes, or for which there are multiple alternate pathways (i.e., locus heterogeneity; see Chapter 3) are unlikely to elicit varying responses due to polymorphisms in a single gene—the phenotype is said to be polygenic. Similarly, in the same sense that multiple genes diminish the importance of any one gene in patient response, multiple polymorphisms within one gene (i.e., allelic heterogeneity; see Chapter 3) tend to lessen the utility of pharmacogenetic studies to optimize patient care. The pharmacogenetics of a therapeutic response can become challenging if the gene responsible for the phenotype has many polymorphisms, all with similar phenotypic outcomes. In such cases, the utility of a single genetic test to successfully guide dosing is small. This problem may be alleviated by newer genomic technologies (see Chapter 4).

The two cases described here are the pharmacogenetics of thiopurines and warfarin. These compounds both have narrow therapeutic windows and phenotypes that are determined by one or two genes. They have both been described repeatedly as classic examples of the important role of pharmacogenetics in explaining variations in patient drug responses. They appear here to elucidate other features of both drugs and genetics that are important in understanding the limitations and complexities of pharmacogenetics in clinical use.

5.2.1 Case 1: Pharmacogenetics of Thiopurine S-Methyltransferase

Thiopurines are among the first-line treatments for childhood acute lymphoblastic leukemia, organ transplant recipients, inflammatory bowel disease, and autoimmune diseases.[13] The enzyme thiopurine S-methyltransferase (TPMT) catalyzes the S-methylation of a number of chemotherapeutic prodrugs, such as 6-mercaptopurine (6-MP), 6-thioguanine (6-TG), and azathioprine. In the cell, 6-MP and azathioprine are converted into thioinosine monophosphate and, ultimately, to thioguanosine monophosphate, which in turn is ultimately converted into cytotoxic nucleotide analogs that inhibit DNA and RNA synthesis (Figure 5-2).

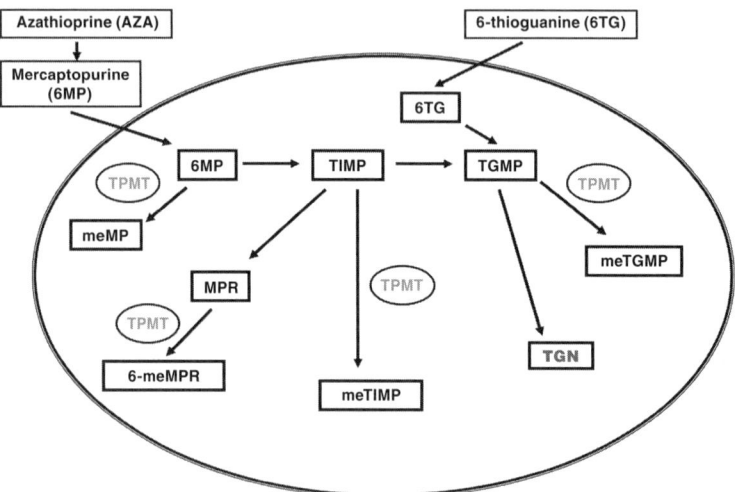

Figure 5-2. The inactive prodrugs 6-MP, 6-TG, and azathioprine are activated by multiple enzymes. After uptake, 6-MP and azathioprine are converted into thioinosine monophosphate (TIMP). 6-TG is converted into thioguanosine monophosphate (TGMP). Thiopurine methyltransferase (TPMT) is the major enzyme involved in the inactivation pathway for all three thiopurines to methylmercaptopurine (meMP), 6-methylmercaptopurine (6-meMPR), methylthioguanosine monophosphate (meTGMP), and methyl-thioinosine monophosphate (meTIMP). Cytotoxic effects of thiopurine drugs occur when cytotoxic nucleotide analogs (TGN) are incorporated into DNA or RNA, stopping synthesis. Adapted from the Pharmacogenomics Knowledge Base (www.pharmgkb.org/index.jsp).

The S-methylation of 6-MP by TPMT is a critical step in the inactivation pathway leading to the ultimate clearance of 6-MP and 6-TG (Figure 5-2). Although a number of pathways are involved in the inactivation of these drugs, one gene product plays a role in all of them (i.e., inactivation is nearly monogenic). High concentrations of the cytotoxic nucleotide analogs have been linked to hematopoietic toxicity and result in low patient tolerance to thiopurine therapies. Loss of the drug inactivation pathways can therefore have important clinical implications.

The genetic role of polymorphisms in TPMT and the resulting enzyme activity was first noted in red blood cells of healthy volunteers.[14] Three groups were identified with high, intermediate, and low enzyme activities. These three groups have now been shown to represent individuals carrying, respectively, no, one, or two variant alleles for TPMT, indicating a clear relationship between phenotype and genotype. Thus, hematopoietic toxicity, the phenotype of interest, is determined in large part by a single gene (i.e., the phenotype is monogenic). Interestingly, another enzyme, xanthine oxidase, also plays a role in inactivation of thiopurines and could confound the role of TPMT in drug response. However, xanthine oxidase is not expressed in hematopoietic tissue, so the polymorphisms that most certainly exist in xanthine oxidase do not affect the phenotypic response of interest, namely, hematopoietic toxicity.

Clinically, polymorphisms in TPMT have been shown to play a role in explaining individual variations in response to thiopurine drug therapy. Low TPMT activity is associated with hematopoietic toxicity in patients treated with standard doses of 6-MP, 6-TG, or azathioprine.[14,15] As of 2008, a total of 29 variant alleles have been identified in TPMT (an example of allelic heterogeneity). Of these variant alleles, four have been most studied with regard to their effects on decreased TPMT activity (Table 5-1) (for a review, see Wang and Weinshilboum[16]). These allelic variants also exhibit geographic differences in their frequencies; *TPMT*2* and *TPMT*3A* are the most common variants in Caucasians, and *TPMT*3B* and *TPMT*3C* are the most common variants in African Americans. The most common variant allele, *TPMT*3A*, is associated with low TPMT activity.[17] Similarly, *TPMT*2* shows loss of catalytic activity, though surprisingly, no difference in *TPMT* messenger RNA concentrations.[18] The low enzyme activity observed for both variants is probably due to a greater rate of degradation.[19] As noted in Chapter 4, this is a classic example in which simply monitoring gene expression levels (via microarrays or quantitative polymerase chain reaction) would not explain the lower enzyme activity.

Although other variants have been identified (*TMPT*4, TMPT*5, TMPT*6*, and *TMPT*7*), they appear to be rare and therefore unlikely to confound the relationship between phenotype and genotype, at least among the populations studied to date. However, the four common variant alleles do not explain all the side effects associated with thiopurine therapy.[20] As is often the problem in many other pharmacogenetic cases, there are other variant alleles for TPMT whose frequencies, and therefore importance, in other populations is yet unknown. The role of environmental factors in patient response is also likely to be important. In a review examining the role of genetic variation in TPMT-mediated adverse drug reactions, van Aken et al.[20] found that 78% of the adverse drug reactions could not be accounted for by the limited number of polymorphisms generally examined in TPMT. These authors point out the need for further studies identifying additional variant alleles in other ethnic groups and for continued careful clinical monitoring of adverse drug reactions.

5.2.2 Case 2: Pharmacogenetics of Warfarin

Warfarin is a commonly prescribed oral anticoagulant for the prevention and treatment of myocardial infarction, ischemic stroke, venous thrombosis, and atrial fibrillation. Warfarin is a very effective antagonist of the vitamin K epoxide reductase complex (*VKORC1*), a critical enzyme in the vitamin K–dependent clotting pathway. Warfarin has a narrow therapeutic window with large interpatient variation. Insufficient drug concentrations may prevent thromboembolism, whereas overdosing increases the risk of bleeding events. Warfarin is delivered as a racemic mixture of the R and S stereoisomers. The stereoisomers are metabolized

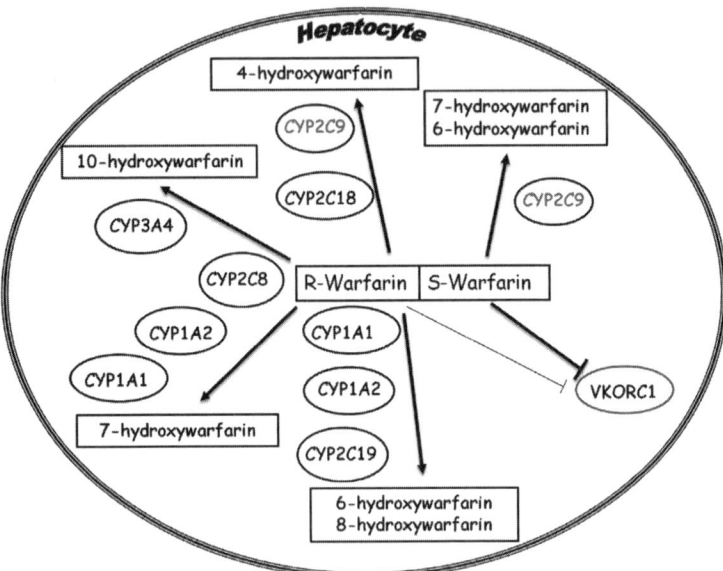

Figure 5-3. Warfarin, a natural product, is given as a racemic mixture of the R and S stereoisomers of the drug. The stereoisomers are metabolized by a number of oxidative phase 1 enzymes, though significantly, the S isomer is metabolized by a single enzyme, *CYP2C9*. In contrast, the less active R-warfarin is metabolized mainly by *CYP3A4* but also *CYP1A1, CYP1A2, CYP2C8, CYP2C9, CYP2C18,* and *CYP2C19*. Warfarin, particularly the S isomer, is a potent inhibitor of vitamin K epoxide reductase complex (*VKORC1*), a critical enzyme in the vitamin K–dependent clotting pathway. Adapted from the Pharmacogenomics Knowledge Base (www.pharmgkb.org/index.jsp).

by different members of the cytochrome P450 phase 1 enzymes. The potency of S-warfarin in inhibiting *VKORC1* exceeds that of the R isomer by three- to fivefold, and the S isomer accounts for 60% to 70% of the anticoagulation response.[21] This is critical in terms of warfarin's pharmacogenetics, because S-warfarin is largely metabolized by a single enzyme (*CYP2C9*) and thus behaves as a monogenic trait (Figure 5-3). In contrast, R-warfarin is metabolized by a number of *CYP* enzymes, mainly *CYP3A4* and, to a lesser degree, *CYP1A1, CYP1A2, CYP2C8, CYP2C9, CYP2C18,* and *CYP2C19*. If R-warfarin were the most active agent, it is unlikely that this drug would be important pharmacogenetically, because so many different genes (polygenic) are involved in its metabolism that no single set of polymorphisms would be useful predictors of therapeutic outcome.

To date, more than 50 variants in *CYP2C9* have been described in human populations, and at least 24 are nonsynonymous substitutions resulting in proteins with altered amino acid sequences. Two variants, *CYP2C9*2* and *CYP2C9*3*, are the most common and most extensively studied (Table 5-2). Because patients with *CYP2C9*2* and/or *CYP2C9*3* variants metabolize warfarin more slowly, traditional

Table 5-1. Major Allelic Variants of TPMT Among Humans*

Allele		mRNA	Protein	Phenotype
*TPMT*2*	rs1800462	G238C	Ala80Pro	100-fold decrease
*TPMT*3B*	rs1800460	G615A	Ala154The	Reduced activity
*TPMT*3C*	rs1142345	A874G	Tyr240Cys	Reduced activity
*TPMT*3A*	—	Includes *3A & *3C		Reduced activity
*TPMT*4*	rs1800584	G → A splice variant		Reduced activity

*Alleles are also used to designate haplotypes, because some of the alleles *3A and *3C are in linkage disequilibrium. *TMPT*3B* is a rare variant and is also in tight linkage disequilibrium with *TMPT*3C*. *TMPT*2*, *TMPT*3A*, and *TMPT*3C* account for over 90% of known variants among humans. Also shown are the database SNP ID#s for each allele.

dosing regimens may lead to bleeding events or longer times to achieve stable drug concentrations versus dose, during which bleeding events may occur. Other polymorphisms that occur at much lower frequencies have not been evaluated. The frequencies of *CYP2C9*2* and *CYP2C9*3* vary considerably among ethnic populations. Among Caucasians, the frequency of *CYP2C9*2* varies from 8% to 20%, and that of *CYP2C9*3* varies from 6% to 10%. Unfortunately, in respect to their utility as general predictors of patient response to warfarin therapy, *CYP2C9*2* and *CYP2C9*3* are largely absent in Asian populations and rare in African American populations, where frequencies range from 1% to 4%. Once again, the distribution of clinically important alleles among human populations is important and limits the universal application of data gathered from one ethnic group. This problem cannot be overstated. The advancements to be derived from pharmacogenetics will depend on the cataloging of all relevant variants in human populations and the development of large-scale genetic screening technologies to identify these alleles.

Yet another issue in the warfarin pharmacogenetics story is locus heterogeneity. The target of warfarin (*VKORC1*) also has variant alleles that affect how patients respond to therapy. This is an example of the pharmacogenetics of drug targets, which, unlike that of genes involved in drug metabolism or drug transport, often result in differences in the pharmacodynamics of response. Mutations in *VKORC1* have been identified in vitamin K deficiency disorders and warfarin resistance. There are at least five important variants for *VKORC1*, as well as numerous less common variants. Fortunately, these variants can be grouped into four haplotypes—chromosomal segments within which the DNA sequence is invariant or constant among most human populations (see linkage and HapMaps, Chapters 3 and 4). These four haplotypes (*VKORC1*1, VKORC1*2, VKORC1*3,* and *VKORC1*4*) include most of the common single-nucleotide polymorphisms (SNPs) that contribute to interpatient variation in warfarin dosing in Caucasians

Table 5-2. Major Allelic Variants of *CYP2C9* and *VKORC1* Among Humans*

Allele		mRNA	Protein	Phenotype
CYP2C9*2	rs1799853	C430T	Arg144Cys	Metabolism reduced 50%
CYP2C9*3	rs1057910	A1075C	Ile359Leu	Metabolism reduced
VKORC1*2	rs9923231	G-3673A C6484T	Promoter in intron	Requires lower dose
VKORC1*3	rs7294	G9041A G3730A	In 3' UTR	Higher warfarin dose

* *VKORC1*2* and *VKORC1*3* designate haplotypes occurring in either promoter region (*2) or in the 3' untranslated region (*3). Also shown are the database SNP ID#s for each allele.

(Table 5-2). *VKORC1*1* is considered the reference sequence and is probably the ancestral haplotype. Individuals with the *VKORC1*2* haplotype (also confusingly referred to as haplotype group A) require lower warfarin doses. This haplotype is common in Asians and Caucasians and rare in African populations.[22] *VKORC1*3* and *VKORC1*4* (referred to as haplotype group B) require a higher warfarin dose. *VKORC1*3* is the most common haplotype in African populations and is also common in Caucasians.

Recently, another *CYP* gene has been identified that has a clinically important impact on the ability of patients to reach stable warfarin dosing.[23] An allele in *CYP4F2* that occurs with moderate frequency in Asians and Caucasians (~30%) but with low frequency in African Americans (7%) results in higher warfarin doses to achieve stable dosing. Thus the warfarin story becomes less perfect and potentially more confusing in its interpretation for patient care. The metabolic role of *CYP4F2* is as yet unknown.

5.3 Summary

These two cases have been highlighted to illustrate some of the complexities that are common to many pharmacogenetic cases. The pharmacogenetic litera- ture contains many examples of confusing or even contradictory studies that arise due to:

- Unknown environmental factors that result in poor outcomes.
- Drugs whose metabolism and/or transport are affected by multiple genes (polygenic) in multiple pathways (locus heterogeneity).
- Clinically important genes that have many rare allelic variants with similar phenotypes (allelic heterogeneity).
- Variation in the frequencies of allelic variants among ethnic groups that mask the role of any one variant.

These issues are common to most gene–drug dynamics and do not preclude the importance of pharmacogenetic studies. They do call for more realistic assessments of the role of genetic testing for the practicing clinician as this field develops. With an understanding of the genomic science underlying patient drug response and the ability to take advantage of resources like the Pharmacogenomics Knowledge Base and the National Center for Biotechnology Information (see Chapter 4), pharmacists and other health care providers will be better prepared both to utilize the genomic data now being collected and to assimilate the new genomic discoveries that are to come.

QUESTIONS

1. It is often the case that "the overall effects of medications are typically not monogenic traits." What is meant by this statement, and what does this mean in terms of phenotypic outcomes?

2. It has been said that "an abnormal drug response that clusters within families provides strong evidence for a genetic cause." Does this provide sufficient proof?

3. Define "locus heterogeneity" and describe how it might cause confusion in the application of pharmacogenetics in the clinic.

4. Define "allelic heterogeneity" and describe how it might cause confusion in the application of pharmacogenetics in the clinic.

5. It is becoming increasingly apparent that an understanding of the pharmacogenetic factors that influence drug metabolism is important in understanding drug response and toxicity, at least for some drugs with narrow therapeutic ranges. However, as discussed, the pharmacogenetic differences in drug metabolism do not fully explain the variability observed in drug response. Describe at least two factors that may contribute to the variability in drug response besides genetics.

References

1. Lindpaintner K. Pharmacogenetics and pharmacogenomics in drug discovery and development: an overview. *Clin Chem Lab Med*. 2003;41:398–410.
2. Kalow W. Human pharmacogenomics: the development of a science. *Hum Genomics*. 2004;1:375–80.
3. Weinshilboum R, Wang L. Pharmacogenomics: bench to bedside. *Nat Rev Drug Discov*. 2004;3:739–48.

4. Efferth T, Volm M. Pharmacogenetics for individualized cancer chemotherapy. *Pharmacol Ther*. 2005;107:155–76.
5. Marzolini C, Paus E, Buclin T, et al. Polymorphisms in human MDR1 (P-glycoprotein): recent advances and clinical relevance. *Clin Pharmacol Ther*. 2004;75:13–33.
6. Nebert DW, Vesell ES. Advances in pharmacogenomics and individualized drug therapy: exciting challenges that lie ahead. *Eur J Pharmacol*. 2004;500:267–80.
7. Nebert DW, Jorge-Nebert L, Vesell ES. Pharmacogenomics and "individualized drug therapy": high expectations and disappointing achievements. *Am J Pharmacogenomics*. 2003;3:361–70.
8. Nebert DW, Vesell ES. Can personalized drug therapy be achieved? A closer look at pharmaco-metabonomics. *Trends Pharmacol Sci*. 2006;27:580–6.
9. Shastry BS. Pharmacogenetics and the concept of individualized medicine. *Pharmaco-genomics J*. 2006;6:16–21.
10. Nebert DW, Zhang G, Vesell ES. From human genetics and genomics to pharmacogenetics and pharmacogenomics: past lessons, future directions. *Drug Metab Rev*. 2008;40:187–224.
11. Transon C, Leeman T, Dayer P. In vitro comparative inhibition profiles of major human drug metabolising cytochrome P450 isozymes (CYP2C9, CYP2D6 and CYP3A4) by HMG-CoA reductase inhibitors. *Eur J Clin Pharmacol*. 1996;50:201–15.
12. Tyndale RF, Droll KP, Sellers EM. Genetically deficient CYP2D6 metabolism provides protection against oral opiate dependence. *Pharmacogenetics*. 1997;7:375–9.
13. Lennard L. The clinical pharmacology of 6-mercaptopurine. *Eur J Clin Pharmacol*. 1992;43:329–39.
14. Weinshilboum RM, Sladek SL. Mercaptopurine pharmacogenetics: monogenic inheritance of erythrocyte thiopurine methyltransferase activity. *Am J Hum Genet*. 1980;32:651–62.
15. Krynetski EY, Evans WE. Pharmacogenetics as a molecular basis for individualized drug therapy: the thiopurine S-methyltransferase paradigm. *Pharm Res*. 1999;16:342–9.
16. Wang L, Weinshilboum R. Thiopurine S-methyltransferase pharmacogenetics: insights, challenges and future directions. *Oncogene*. 2006;25:1629–38.
17. Tai H-L, Krynetski EY, Schuetz EG, et al. Enhanced proteolysis of thiopurine S-methyl-transferase (TPMT) encoded by mutant alleles in humans (TPMT*3A, TPMT*2): mechanisms for the genetic polymorphism of TPMT activity. *Proc Natl Acad Sci USA*. 1997;94:6444–9.
18. Krynetski EY, Fessing MY, Yates CR, et al. Promoter and intronic sequences of the human thiopurine S-methyltransferase (TPMT) gene isolated from a human PAC1 genomic library. *Pharm Res*. 1997;14:1672–8.
19. Wang L, Sullivan W, Toft D, et al. Thiopurine S-methyltransferase pharmacogenetics: chaperone protein association and allozyme degradation. *Pharmacogenetics*. 2003;13:555–64.
20. van Aken J, Schmedders M, Feuerstein G, et al. Prospects and limits of pharmacogenetics: the thiopurine methyl transferase (TPMT) experience. *Am J Pharmacogenomics*. 2003;3:149–55.
21. Yin T, Miyata T. Warfarin dose and the pharmacogenomics of CYP2C9 and VKORC1—rationale and perspectives. *Thromb Res*. 2007;120:1–10.
22. Geisen C, Watzka M, Sittinger K, et al. VKORC1 haplotypes and their impact on the inter-individual and inter-ethnical variability of oral anticoagulation. *Thromb Haemost*. 2005;94:773–9.
23. Caldwell MD, Awad T, Johnson JA, et al. CYP4F2 genetic variant alters required warfarin dose. *Blood*. 2008;111:4106–12.

CHAPTER 6

Ethical Challenges and Opportunities in Pharmacogenetics and Pharmacogenomics

LEARNING OUTCOMES:

At the end of this chapter, you should be able to:

1. Define relevant terms and identify the various individuals and groups involved in the ethical issues related to pharmacogenetics and pharmacogenomics.

2. Discuss, compare, and contrast ethical considerations and issues related to pharmacogenomics and pharmacogenetics for selected individuals and groups in the contemporary health care system.

As the field of pharmacogenomics gains utility in clinical practice, primary care and specialty physicians, genetic counselors, nurses, and pharmacists will play a role in interpreting and communicating pharmacogenomic information to patients. Given their education and expertise in the interpatient variability of drug metabolism, disposition, and pharmacodynamics, pharmacists are well placed to assume the leading role as the "learned intermediary" to consumers and patients with regard to pharmacogenomic information and how it can optimize and enhance their health care.

With the expanding impact of pharmacogenetics and pharmacogenomics on patient care in all settings, pharmacists and other health care professionals will face ethical issues that represent novel challenges. Pharmacists will be particularly challenged to consider various pharmacogenomic ethical issues and dilemmas because of their interactions with patients through individualized counseling, medication therapy management programs, medication reconciliation programs, transitional care programs, and consultation activities in community and institutional settings. Furthermore, the easy accessibility of pharmacists in a variety of professional settings is likely to result in patients and customers seeking help to understand these concepts and issues. This will be extremely important given the increasing amounts of genetic information and services, some of questionable validity, found on the Internet.

This chapter provides a brief, general overview of the key terms and issues, challenges, and opportunities associated with ethics and ethical problem solving as they relate to pharmacogenomics, pharmacogenetics, and personalized medicine. These ethical challenges and opportunities will have impacts on a wide variety of individuals and groups (Table 6-1). New pharmacogenomic ethical challenges will arise in the years to come, necessitating pharmacists and other health care professionals to utilize various models of ethical problem solving and decision making in specific circumstances.[1-3]

6.1 Medical Ethics and Its Relation to Personalized Medicine and Clinical Practice

In 1979, the U.S. Department of Health, Education, and Welfare (now the Department of Health and Human Services) released a report titled *Ethical Principles and Guidelines for the Protection of Human Subjects of Research*, commonly referred to as the Belmont report. This document provides an important framework within which to consider the basic principles of ethics in relation to pharmacogenomics.[4] The Belmont report outlined the basic principles—specifically, respect for individuals, beneficence, and justice—that should guide the ethical treatment of patients in research and clinical practice. This section discusses the latter two of these concepts.

Table 6-1. Individuals and Groups Potentially Engaged in Pharmacogenomics Ethical Issues and Challenges

Individuals	Groups
• Health care professionals	• Health and benefits management organizations
• Health professions educators	• Health care organizations (hospitals, clinics)
• Health-related administrators	• Community pharmacies
• Patients	• Professional organizations
• Legislators	• Unions
• Judges	• Religious organizations
• Scientists	• Charities
• Regulators	• Pharmaceutical industry
• Union officials	• Local, state, and national legal and justice system
• Clergy	• Local, state, and national government bodies
	• Local, state, and national policy-making boards

6.1.1 The Belmont Report

In medical ethics, the concept of **justice** holds that all people—regardless of age, ethnicity, gender, national origin, race, religion, sexual orientation, social status, economic status, or any other characteristic or attribute—should be treated equally and should receive the same quality of medical care. The Belmont report defines justice as the precept that persons who are equals in all medically relevant respects should receive equal medical care.[4] **Injustice** occurs when individuals are treated differently on the basis of non–medically relevant characteristics, such as age, race, gender, and so forth.

For example, the treatment of an adult patient can be expected to differ from that of a child. This is not unjust, because age is a medically relevant characteristic that justifies differences in these patients' medical care. In contrast, injustice occurs when a patient is not afforded the same access to or quality of medical care because of his or her race or socioeconomic status. Another example of how the concept of justice applies to medical care is the difference in the ability of an adult parent versus that of a minor child to participate in the decision-making process with regard to his or her own health care. Parents are afforded the power to make decisions about their children's care, because of their age and their guardianship status in relation to their children. This difference, therefore, does not constitute injustice. However, if an adult is disallowed or discouraged from participating in his or her own care because of one of the previously named characteristics (race, sex, etc.), this constitutes injustice.

Western medical ethics, which has its origins in **Hippocratic ethics**, is based on the premise that the health care professional should, "first, do no harm" to the patient. Patients expect pharmacists and other health care professionals to demonstrate **beneficence** in their care. Beneficence is also one of the basic ethical principles outlined in the Belmont report.[4]

Justice:
In medical ethics, the precept that all people, regardless of age, ethnicity, gender, national origin, race, religion, sexual orientation, social or economic status, or any other attribute or characteristic, should be treated equally and should receive the same quality of medical care.

Injustice:
In medical ethics, the lack of fairness or justice that occurs when individuals who are equal in all medically relevant respects are treated differently on the basis of characteristics such as age, race, gender, or other non–medically relevant factors.

Hippocratic ethics:
A system of medical ethics based on the so-called Oath of Hippocrates and asserting six fundamental principles for the practice of medicine: (1) respect for one's teachers and consideration toward one's colleagues, (2) the command "first, do no harm," (3) respect for the patient's life, (4) restriction of one's practice to one's field of expertise, (5) refraining from abuse of privilege, and (6) respect for the patient's privacy.

The Belmont report asserts that clinicians should always be guided by two overarching principles when acting in the interests of their patients: they should (1) do no harm and (2) maximize the potential benefits while minimizing any possible harms of the care they provide. For the pharmacist, beneficence can be defined as working for the benefit of patients and assisting them in managing their medications—specifically, weighing the therapeutic benefits of a medication against its potential adverse effects. Pharmacists must also work with other health care professionals to prevent harm to their patients—that is, to practice **nonmaleficence**—by preventing the ineffective or inappropriate use of medications and to maximize their benefits through the knowledgeable application of current therapies.

> **Beneficence:**
> An ethical principle asserting that a health care provider must always act in the best interests of the patient, both by improving the patient's well-being and by preventing harm to the patient.
>
> **Nonmaleficence:**
> An ethical principle holding that the health care provider should refrain from a course of action that would bring harm to the patient.

6.1.2 Personalized Medicine

The concept of personalized medicine, or tailoring therapy to a specific patient, presents challenges to the existing process for developing new prescription therapies. Currently, this process is based on the premise that an approved therapeutic agent is generally safe but carries the potential risk of side effects or adverse effects for some of the population. Personalized medicine is at odds with a major starting point in our current concept of "one-size-fits-all" drug development and utilization. Drug therapies are currently developed based on a disease state or condition, with the expectation that they will be targeted to a large population. It is understood that the drug, though safe, may not benefit all patients equally.[5–9] In contrast, personalized medicine seeks to identify patients who, because of their genetic background or individual disease state, are most likely to have a successful therapeutic outcome and least likely to experience adverse effects from the therapy.

The potential for injustice and maleficence (potential for doing harm) for a patient arises when a practitioner's clinical behavior leads to differences in his or her treatment of patients with outwardly similar disease states or conditions. For example, a pharmacist or health care provider may justly recommend the use of a medication on the basis of a patient's pharmacogenetic profile. Alternatively, the same pharmacist or health care provider may be reluctant, on the basis of the scientific literature, to use a particular therapeutic agent in a patient with the same disease state or condition because this patient's pharmacogenetic profile suggests that the medication will have a limited potential benefit and therefore may not result in the desired therapeutic outcome. However, our limited knowledge of the pharmacogenomic elements of most drugs, combined with specific patient factors that may be unknown, is such that, in many cases, the pharmacist or health care provider cannot rule out the possibility that a specific pharmacotherapeutic agent will or will not result in a successful outcome for a particular patient. Injustice and potential harm could result for this patient if he or she is not given access to an available treatment because of the professional opinion (behavior) of the pharmacist or health care provider.

Injustice can also arise if the organization managing a patient's pharmacy benefits restricts access to therapeutic agents by selecting treatments on the basis of pharmacogenetic criteria, without considering the complexities of genomic information. In this case, ethical and legal questions will arise if the organization denies a medication because the patient does not fit the pharmacogenetic profile.[10–12] These questions also pertain to whether the health care provider can justify the use of the medication or has the autonomy in his or her practice to prescribe or administer a given therapeutic agent.[13] Due to their extensive education and clinical experience in medication use, pharmacokinetics, and pharmacodynamics to promote successful therapeutic outcomes, pharmacists can play a key leadership role in these decision-making processes in clinics, health care organizations, and health benefits organizations.[14] A basic understanding of genomic principles is essentially no less than an understanding of pharmacokinetic principles for successful pharmacy practitioners in the health care environment, both now and in the future.

6.2 Ethical Implications for Pharmacogenomics and Pharmacogenetics

6.2.1 Patients

The assumption underlying our current use of prescription medications continues to be a trial-and-error process whose goal is to select the most appropriate therapy on the basis of the scientific literature, current practice guidelines, and the individual patient. The goal is to balance the risk against the benefits associated with the use of therapeutic agents to manage a disease condition or state, realizing that some patients will be nonresponsive and/or have adverse reactions. One issue to be addressed is to what extent we, as a society, are willing to distribute the risk that is created among different genetic groups as we utilize pharmacogenomics and pharmacogenetics in the drug discovery and development process.[15–17] For example, does a patient's characterization in a particular genetic group justify the denial of access by this individual to a therapeutic agent during clinical testing?[16] A second dilemma is how society and health care professionals will deal with a genetically defined subgroup and the **stigmatization** (i.e., negative characterization of patients on the basis of their genetic profile) and/or social discrimination that may be associated with the results of genetic testing.[10–11,18]

Patient **autonomy**, or the ability of a patient to independently decide upon his or her own course of action, will play a controversial role in health care as pharmacogenetic differences are better identified and characterized. For patients to make informed and independent decisions about their health care, it is key for them to have knowledge about their health conditions and possible treatments. Pharmacists and health care providers play an important role to assist patients by providing critical information for this decision-making process. The roles of pharmacists and other health care providers will become particularly crucial in the area of genetics and pharmacogenomics, given the complexity and nuances of the field. The critical next question becomes to determine the responsibility of a health care provider to inform a patient about any pharmacogenetic factors that may affect the efficacy or toxicity of a drug. This is a particularly troublesome question because, in many cases, pharmacogenomic information is unlikely to be definitive at a specific time, although it may become more defined with advances in clinical sciences and treatment.

> **Stigmatization:**
> The negative characterization or categorization of patients on the basis of their genetic profile or other characteristics, such as age, race, or gender.
>
> **Autonomy:**
> The ability of a person to act independently and to make meaningful decisions without external coercion or controlling interference.

A number of other ethical and legal issues will arise with the increasing availability of pharmacogenetic information. It will be essential to characterize the responsibilities of the health care provider and the health care organization or benefits manager, both ethically and legally, if a patient or his or her legal guardian refuses to utilize a medication even though, pharmacogenetically, it may be known to have a beneficial effect.[16] In addition, pharmacists and other health care providers will still face the ethical issues associated with treating patients with diminished autonomy and the extent to which they must protect these patients.[5] Ethical decision-making processes will also be needed to determine what genetic information should be used to optimize patient care, who should make decisions about a patient's care, and to what extent genetic information should be shared among family members and other parties.[7,11]

Patient confidentiality and privacy has always been one of the foremost elements of patient care in any health care setting. Policy makers at all levels will be faced with the development of legislation, rules, and guidelines to address a number of questions related to patient confidentiality, such as:[19,20]

- Who should have access to a patient's pharmacogenetic or genetic profile?
- How should this information be used to minimize toxicity or enhance the cost-effectiveness of a pharmacotherapeutic management plan?

▪ What are the requirements for storing patients' confidential information?

▪ What safeguards or precautions must be in place when confidential information is shared between health care providers and health care organizations?

Information about a patient's genetic makeup is unlike the results of physical assessment and clinical testing (e.g., blood pressure, blood glucose, lipid levels) in that this information applies not only to a given individual but may also provide insight into other related individuals and may be linked to potential disease states. It will also need to be decided, at all levels of the decision-making process in our health care system, whether family members should be informed of potential genetic factors that could affect their subsequent care. The impact of pharmacogenetic information on pharmacy practice has been elucidated by Rothstein and Epps, who reminded us that: "As information regarding the genotype of an individual becomes increasingly important to safe prescription and dosage, pharmacists might be charged with greater knowledge of their customers' genetic information than they now require."[21]

6.2.2 Pharmacists and Other Health Care Professionals

If pharmacists and other health care professionals are expected to utilize pharmacogenetic information in the care of their patients, many issues will need to be addressed within health care organizations. It will remain essential for a patient's genetic information to remain as secure as other elements of his or her personal health care record. The intersection of bioethics with business ethics and advances in bioinformatics and databases will provide new ethical challenges in how genetic and pharmacogenomic information is utilized and shared between key parties.[22] Technological advancements in electronic medical records will need to be developed that will allow this genetic information, like other aspects of personal health information, to be confidentially and securely stored and transmitted. Pharmacogenetic testing, as it relates to the care of patients, will most certainly present challenges related to patient confidentiality, the personal health record, and to what extent and how this information is shared between organizations and with patients and their families. Pharmacists can and must play a key role in these developments and the implementation of these new data management and bioinformatics systems to ensure optimized care of their patients with available pharmacogenomic testing and information.

Legal issues may arise if health care professionals are not provided with relevant pharmacogenomic results as they develop care plans or if they do not share this critical genetic information as patients utilize services from different health care organizations. A health care professional who has access to this information but fails to utilize it in a patient's care will also be at risk for potential liability. Not understanding the genetic information in a patient's medical record will not be a suitable defense for health care providers or organizations who find themselves in such legal situations.

Issues regarding sharing and availability of patient information are certainly not new in contemporary health care. As noted earlier, genetic information, unlike the results of other medical tests, may be correlated with risks for other diseases other than those intended by genetic screening. These risks may require the elucidation of additional legal rights and responsibilities of pharmacists and other health care professionals as they relate to liability issues.[13] Furthermore, expectations for standards of care will be modified as pharmacogenetic differences identify key considerations in the development of patient care plans.

Contemporary clinical practice is focused on current standards of care, which do not at this stage require specific genetic testing as a component of managing patient care. However, with continued advances in pharmacogenetics, it will most likely be essential to identify patients who may either benefit from or experience adverse effects related to a therapeutic agent based upon genetic testing.[18,20] The future standard of care will certainly require genetic testing prior to the prescribing of a specific drug. What complicates the use of pharmacogenetic testing is that a simple relationship does not exist between such testing and one gene, one disease, or one therapy, and validating a particular genetic test will be a complicated and tedious process. Furthermore, pharmacogenetic testing is complex, owing to issues related to clinical validity and utility, test interpretation, quality assurance for the development of the test combined with the necessary education of pharmacists and other health care providers related to the test timing, informed consent, and genetic counseling, as well as the secondary use of this information.[8,23–26]

A drug's approved labeling may make it necessary for patients to be tested before being prescribed the drug to ensure its efficacy and safety. Any such labeling would need to be reflected in clinical guidelines from professional health care groups and organizations. Drug utilization reviews would also need to reflect these requirements for appropriate pharmacogenetic testing. One can envision situations in which a pharmacist or other health care provider becomes liable if he or she is unaware of the requirement for pharmacogenetic tests or is unable to interpret the results when establishing a patient care plan. The availability of a patient's pharmacogenetic information that is related to the use of a specific medication will place greater responsibilities on pharmacists as they participate in medication management, collaborative drug therapy management, and transitional care. Pharmacists will also need to remember the importance of the secondary information that may arise from these pharmacogenetic tests. Specifically, to what extent will pharmacogenetic results be related to the treatment of other diseases or the use of other drugs, and what is the responsibility of the pharmacist or other health care professional to be aware of these relationships and their impact on patient care?[25]

Pharmacogenetics will also present challenges to contemporary health care practice with regard to off-label prescribing. Pharmacists and other health care professionals, as well as health care organizations, could be liable if adverse drug reactions occur in patients who are subsequently identified with an incompatible genotype, resulting in what could then be considered off-label prescribing. Issues requiring answers will include (1) the extent to which health care organizations should be responsible for notifying patients who are now known to be at risk for this adverse drug reaction; (2) the extent to which these health care professionals should be liable, considering that they were practicing according to the current standard of care; and (3) whether there should be limits on the extent to which pharmacists, health care professionals, and health care organizations or pharmacies are liable in these situations.

6.2.3 Health Care Organizations, Insurance, and Pharmacies

Health care organizations that manage patient care issues will face similar challenges with advancements in pharmacogenomics and pharmacogenetics.[5,12,18] Cost-benefit analysis will be the major issue. These organizations may face questions as to who should pay for the increased cost associated with required pharmacogenetic tests. Should this be the patient's responsibility or part of the standard of care whose cost is shared by all subscribers of a health care plan? What if a patient refuses to pay for such a test? Would it be appropriate for a health care organization to deny the patient access to the drug if it has the results of this test on file? Alternatively, can a health benefits organization deny a patient a therapeutic agent if he or she does not fit the pharmacogenetic profile for the drug? Drug selection, utilization, and the processes of drug review in our health care organizations will need to evaluate any required pharmacogenomic testing procedures related to specific therapeutic agents and diseases during its formulary review process. For example, selection of appropriate chemotherapeutic agents will be based on tumor genetics, and optimization of therapeutic benefits will be based on specific pharmacogenomics testing that would contribute to therapeutic effectiveness or minimize toxicity based on pharmacokinetics. Pharmacists in all settings will be required to enhance their knowledge and skills in pharmacogenetic testing and evaluation as applied to clinical care, because the amount of available information will rapidly increase with the development of new and easier technologies to quickly and accurately identify an individual's pharmacogenetic profile.

Pharmacogenomics and pharmacogenetics must be an area for which pharmacists will need to demonstrate competence through continuing professional development programs, and this competence must be based on a strong foundation in genetics and genomics. Continuing professional development, similar to what is expected for other areas, such as law or HIV/AIDS, will need to be demonstrated on a yearly basis, given the growth of contemporary practice. Pharmacists must also look forward to how they will need to advance the scope of their professional practices to facilitate patient care and incorporate their knowledge of pharmacogenomics, pharmacogenetics, and pharmacogenetic testing. Pharmacy practice organizations must educate legislators and legislative bodies on the expertise that pharmacists can bring to optimizing patient care and safety with pharmacogenetic and pharmacogenomic information.

Although any pharmacist is at risk and potentially liable for not obtaining or using available pharmacogenomic and pharmacogenetic information in the care of patients, community pharmacists may be at greater risk for liability, particularly for therapeutic agents that require pharmacogenetic testing prior to being prescribed or dispensed. Community pharmacists may not have ready access to this information because of the lack of standardized patient medical records and differences among health care organizations and companies. Obtaining this information directly from their patients during counseling will be an additional challenge. Community pharmacists and their professional organizations must be key leaders and advocates for access to this key pharmacogenetic information as it relates to the care of their patients. Ethical and legal issues will arise as to the responsibility of health care providers and health care organizations to provide a given medication to a patient when they do not have the necessary access to pharmacogenetic testing results to evaluate the drug's efficacy and safety.

6.2.4 The Pharmaceutical Industry and Drug Development and Discovery

Advances in pharmacogenomics and pharmacogenetics will present great opportunities and challenges in the discovery and development of new drugs.[5,7–9,16–18,27] The current economic reality is based on developing pharmaceuticals that are safe and effective for large patient populations and offsetting the cost of development with the profits from selling to a larger market. With the increasing availability of pharmacogenomic and pharmacogenetic technologies, this reality is likely to be fragmented.[5] It may become necessary to develop alternative economic models to account for a reality in which the size of the patient group, which is classified according to pharmacogenomic differences or disease subtypes, may not necessarily provide the requisite profit margin for the development of a pharmaceutical agent within the existing paradigm of drug development and approval. The identification of pharmacogenetic differences could easily be the key to advancing the efficacy and safety of what were once considered therapeutic "orphan" drugs. Translational and clinical research could identity specific patient groups for which existing therapeutic agents could be revitalized once it is shown that they are particularly effective and/or less toxic for those patients' disease states or conditions. Furthermore, identifying pharmacogenetic differences among patients could restore drug candidates whose development had been formally cancelled because of challenges relating to pharmacokinetic, pharmacodynamic, and toxicity issues, and that can now be beneficial to this smaller, specific patient group. Alternatively, the availability of pharmacogenetic testing could lead to more orphan drugs for which the clinical benefit, based on pharmacogenetic differences, may be limited to such a small patient population that the cost of developing the drug would clearly exceed any profits that could be made from use. The question will then arise as to who should pay for the development of therapeutic agents intended for a limited potential patient population. Should the market bear the costs of developing these drugs, which will then be made available to health care companies or to patients who can afford them? This in turn will lead to questions about the role of government agencies, such as the National Institutes of Health and the Food and Drug Administration, in working with the pharmaceutical industry in the development and approval of new therapeutic entities or the revitalization of existing therapies for limited patient populations.

6.2.5 Clinical Trial Design and Pharmacoepidemiology

Given their scientific and clinical expertise, pharmacists have a great professional opportunity to engage in clinical trial design, clinical trial access issues, and pharmacoepidemiology when there are pharmacogenomic and pharmacogenetic considerations.[28-32] It may be necessary for clinical trial designs to incorporate pharmacogenomic considerations into their inclusion and exclusion criteria. It can be questioned whether pharmacogenomic considerations are appropriate to include in these criteria, given that this might bias the study results to the benefit of the study sponsor or might significantly limit the number of patients engaged in the study. For example, during the clinical testing phase, a drug sponsor could decide that the study patients are difficult to treat and may discontinue the study. Alternatively, the drug sponsor may discover that the cost-profit ratio is not sufficient to warrant the expense of continuing the study, even though some patients have experienced significant therapeutic benefit. Such situations could lead to more orphan drugs or diseases. The question then becomes what is the responsibility of drug manufacturers or the federal government to ensure that these new drugs are available to patients who could benefit from them.

Legislation will also be needed to determine whether a drug sponsor or corporate partner should have the right to "bundle" the use of a genetic test for a specific polymorphism with a particular drug, especially if the sponsor incurred the costs to develop the test and if this might prevent or limit lawsuits arising from inappropriate clinical use. The drug approval process may need to be modified and expedited to include the appropriate validation of such genetic tests as essential to the use of a therapeutic agent. It will also be critical to consider who will own the results of these tests—whether it is the company, the health care organization, or the patient. Furthermore, to what extent should a patient's family members have a right to this information if it could have an impact on their health and health care decision-making process? Decisions will need to be made concerning who will be responsible for ensuring that these results are shared with other health care providers or organizations, especially when a patient may have multiple care providers in various organizations and locations.

The above scenarios, though challenging, may be simple when one considers the possibility of **stratification**—that even within a population, polymorphisms of pharmacological and therapeutic importance may vary in frequency. Clinical trials will need to incorporate ways to prevent stigmatization of patients in the study approval process and informed consent. It is possible that a clinical study investigating a specific polymorphism may hold the potential for stigmatization of the study subjects, resulting in

> **Stratification:**
> In a clinical trial, the process of classifying individuals on the basis of distinguishing characteristics, such as age, sex, race, lifestyle, socioeconomic status, and genetic profile.

some patients being less willing to participate in clinical trials lest this information become available to other parties or organizations. An individual's willingness to participate in a clinical study based on pharmacogenetic characteristics could be limited by simply being in a room with other patients. The possible consequences of such patient stratification and stigmatization would include smaller clinical trials, an increased risk of not detecting adverse side effects in a larger population, and the approval of a therapeutic agent with less information about its possible side effects.[18] Pharmacists should play a key role in the development of policies, procedures, and legislation that lead to a drug approval process that will ensure the availability of safe and effective drugs for their patients.

New issues will arise as to the importance of inclusion and exclusion criteria related to race and ethnicity as pharmacogenetic testing suggests better indices for participants in clinical trials.[28–30] It will be questionable whether to include race as a patient characteristic in the design of clinical trials, because racial features are not necessarily useful predictors of a patient's underlying genetics (Figure 6-1). Additional considerations may arise as to the potential impact of clinical studies involving children, confidentiality, and biobanking of clinical samples from these studies.[31] Many of our concepts with regard to pharmacoepidemiology and pharmacoepidemiological research will certainly be modified with the increased

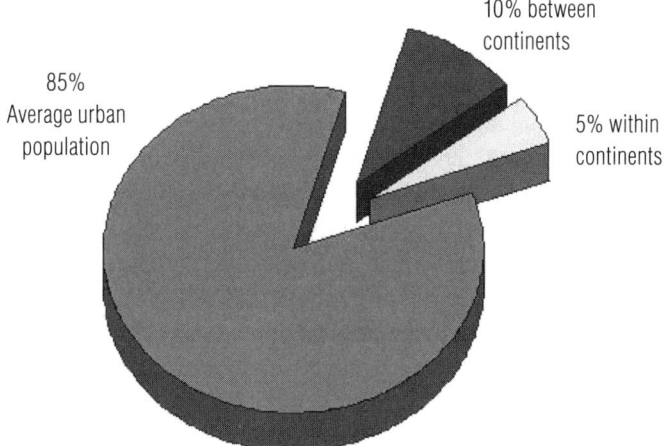

Figure 6-1. Geographic basis of genetic diversity in human populations. The vast majority of the genetic differences among individuals can be found within the average urban population. Very little of the genetic diversity among humans is found between continents. Because racial features are correlated with continents of origin, very little genetic information can be predicated upon racial features alone. Racial features are not a useful predictor of an individual's genetic makeup.

availability of tissue banks that are linked to population databases, as well as the ethical issues of patient informed consent, confidentiality, and the benefits that arise from the prevention and treatment of diseases.[32-34]

Pharmacists must also play a role in what could be changes in postmarket surveillance and public policy. As our knowledge of critical pharmacogenetic differences and their relationship to successful drug therapy improves, the requirements for postmarket surveillance are likely to be modified by the Food and Drug Administration to account for adverse reactions. Pharmacists and other health care providers may be responsible for increasing the confidentiality of their patient records with respect to the genetic aspects of patient care. It will be important to consider to what extent it will be incumbent upon care providers to promote and provide appropriate testing as new testing becomes available in standard clinical practice. All of these elements are likely to have an impact on health policy decisions and scope of practice areas for health care professionals at the state and national levels.

QUESTIONS

1. Discuss the potential for justice or injustice with respect to a new pharmacogenomics test that is suggested to be predictive for a successful therapeutic outcome in the treatment for prostate cancer, but is recommended only to patients of African American descent.

2. In your role as a community consultant pharmacist, you have received the results from a new genetic test indicating that one of your patients may be at risk for endometrial cancer and should immediately be placed on new preventive treatment. You will be seeing this patient shortly to counsel her on future options. You are also a good friend of the patient's family and have also counseled her two sisters in the past. As a good friend of the family, you know the parents well and are aware of the fact that, although the patient is very close to her older sister, she has not spoken with her younger sister for over 5 years. What is your responsibility as a pharmacist to this patient, to her sisters, and to this family?

3. You are the chair of the hospital drug utilization review board. Your hospital medical board of directors has recently approved a new, but controversial, policy developed by the drug utilization review board to not recommend a new pharmacogenetic test that could be useful in determining which patients will require a larger dose when given a new antibiotic for treating community-acquired methicillin-resistant *Staphylococcus aureus*. The board members approved this policy because the scientific literature (in their evaluation) is not definitive and they believe that it would be best to administer the drug without any pharmacogenetic testing at the lower dose to save money. However, a recent program on National Public Radio has received much press on the benefits of this test, and you are receiving calls from the local media on why your hospital has made this decision. What are the key issues you need to consider in preparing for this interview from the local media?

4. Identify the key issues that pharmacists and other health care providers in managed care organizations will need to address with regard to standards of care, responsibilities, rights, and liabilities of health care providers, as well as patient rights.

5. Discuss the positive and negative potential impacts of stratification in the design of clinical trials.

6. What is meant by the term "therapeutic orphans" in regard to pharmacogenomically based drug development?

References

1. Posey LM. *Pharmacy: An Introduction to the Profession*. 2nd ed. Washington, DC: American Pharmacists Association; 2009.
2. Veatch R, Haddad A. *Case Studies in Pharmacy Ethics*. 2nd ed. New York: Oxford University Press; 2008.
3. Hundert EM. A model for ethical problem solving in Medicine, with practical applications. *Focus*. 2003;1:427–35.
4. National Commission for the Protection of Human Subjects of Biomedical and Behavioral Research. *The Belmont Report: Ethical Principles and Guidelines for the Protection of Human Subjects of Research*. Washington, DC: U.S. Department of Health, Education, and Welfare; 1979. http://www.hhs.gov/ohrp/humansubjects/guidance/belmont.htm.
5. Morley KI, Hall WD. Using pharmacogenetic and pharmacogenomics in the treatment of psychiatric disorders: some ethical and economic considerations. *J Mol Med*. 2004;82:21–30.
6. Peters EJ. Eyes on the prize: bringing individualized therapy from the bedside to clinical practice. *Pharmacogenomics*. 2007;8(10):1295–8.
7. Peterson-Iyer K. Pharmacogenomics, ethics and public policy. *Kennedy Inst Ethics J*. 2008;18(1):35–56.
8. Breckenridge A, Lindpaintner K, Lipton P, et al. Pharmacogenomics: ethical problems and solutions. *Nature*. 2004;5:676–80.
9. Issa AM. Ethical perspectives on pharmacogenomic profiling in the drug development process. *Nature*. 2002;1:300–8.
10. Evans JP. Health care in the age of genetic medicine. *JAMA*. 2007;298(22):2670–2.
11. Bereano P. Does genetic research threaten our civil liberties? *ActionBioscience*. 2000. http://www.actionbioscience.org/genomic/bereano.html.
12. Clayton EW. Ethical, legal and social implications of genomic medicine. *N Engl J Med*. 2003;349:562–9.
13. Evans BJ. Finding a liability-free space in which personalized medicine can bloom. *Clin Pharmacol Ther*. 2007;82(4):461–5.
14. Newton R, Lithgow J, Po ALW, et al. *How Will Pharmacogenetics Impact on Pharmacy Practice? Pharmacists' Views and Education Priorities*. Birmingham, UK: National Genetics Education and Development Center and Royal Pharmaceutical Society of Great Britain; 2007. http://www.geneticseducation.nhs.uk/media/16739/Pharmacogenetics.pdf.
15. Wetz DC. Ethical, social and legal issues in pharmacogenomics. *Pharmacogenomics J*. 2003;3:194–6.
16. Barash CI. Ethical issues in pharmacogenomics. *ActionBioscience*. 2001. http://www.actionbioscience.org/genomic/barash.html.
17. Issa AM. Ethical considerations in clinical pharmacogenomics research. *Trends Pharmacol Sci*. 2000;21(7):247–9.

18. Garrison LP Jr, Carlson RJ, Carlson JJ, et al. A review of public policy issues in promoting the development and commercialization of pharmacogenomic applications: challenges and implications. *Drug Metab Rev.* 2008;40:377–401.
19. Anderlik MR, Rothstein MA. Privacy and confidentiality of genetic information: what are the rules for the new science? *Annu Rev Genomics Hum Genet.* 2001;2:401–33.
20. Robertson JA. Consent and privacy in pharmacogenetic testing. *Nat Genet.* 2001;28:207–9.
21. Rothstein MA, Epps PG. Ethical and legal implications of pharmacogenomics. *Nat Rev Genet.* 2001;2:228–31.
22. Goodman KW, Cava A. Bioethics, business ethics, and science of bioinformatics and the future of healthcare. *Camb Q Healthc Ethics.* 2008;17:361–72.
23. Haga SB, Burke W. Pharmacogenetic testing: not as simple as it seems. *Genet Med.* 2008;10(6):391–5.
24. Pendergast MK. Regulatory agency considerations of pharmacogenomics. *Exp Biol Med.* 2008;233:1498–1503.
25. Netzer C, Biller-Andorno N. Pharmacogenetic testing, informed consent and the problem of secondary information. *Bioethics.* 2004;18(4):344–60.
26. Koo SH, Lee EJD. Pharmacogenetics approach to therapeutics. *Clin Exp Pharmacol Physiol.* 2006;33:525–32.
27. March R, Cheeseman K, Doherty M. Pharmacogenetics—legal, ethical and regulatory considerations. *Pharmacogenomics.* 2001;2(4):317–27.
28. Lee SS-J. Racializing drug design: implications of pharmacogenomics for health disparities. *Health Policy Ethics.* 2005;95(12):2133–8.
29. FitzGerald KT. Ethics at the intersection of pharmacoethnicity. *Nature.* 2008;84(3):424–6.
30. Holm S. Pharmacogenetics, race and global injustice. *Dev World Bioeth.* 2008;8(2):82–8.
31. Avard D, Silverstein T, Sillon G, et al. Researchers' perceptions of the ethical implications of pharmacogenomics research with children. *Public Health Genomics.* 2009;12:191–201.
32. Jones JK. Pharmacogenetics and pharmacoepidemiology. *Pharmacoepidemiol Drug Saf.* 2001;10:457–61.
33. deMontgolfier S, Moutel G, Duchange N, et al. Ethical reflections on pharmacogenetics and DNA banking in a cohort of HIV-infected patients. *Pharmacogenetics.* 2002;12(9):667–75.
34. Williams G, Schroeder D. Human genetic banking: altruism, benefit and consent. *N Genetics Soc.* 2004;23(1):89–103.

Glossary

acetylation: A reaction that introduces an acetyl functional group into a chemical compound. Most proteins are modified by acetylation.

allelic heterogeneity: Case where a given phenotype is caused independently by different alleles within the same gene. When the same phenotype is caused independently by two or more distinct genes, it is called locus heterogeneity.

autonomy: The ability of a person to act independently and to make meaningful decisions without external coercion or controlling interference.

autosomal: Genes or loci that reside on any chromosome other than the sex chromosomes (i.e., the X and Y chromosomes).

base coverage: The average number of reads per nucleotide. For example, eightfold base coverage means that, for a given DNA sequence, each base has been independently sequenced eight times. Base coverage, and therefore sequencing effort, will decrease as the quality of sequence data and read length increase.

beneficence: An ethical principle asserting that a health care provider must always act in the best interests of the patient, both by improving the patient's well-being and by preventing harm to the patient.

chimeric: An organism that contains genetic material from a different species.

co-dominant: A common departure from simple mendelian patterns of inheritance occurs when the products of both alleles are detectable in the individual.

codon: A set of three adjacent nucleotides.

consensus sequence: A sequence that indicates the most abundant residues (e.g., nucleotides) at each position, based on multiple sequences. For nucleotide sequences, "N" stands for any base, "Y" represents any pyrimidine base, and "R" indicates any purine base.

diploid: One set of chromosomes in an organism or cell containing two sets of chromosomes (paternal and maternal) is called haploid. If both sets of chromosomes are present, the case for most animal cells except the gametes, the cell is said to be diploid.

DNA polymerase: The enzyme that copies one DNA strand using the complementary strand as a template.

enhancers: Recognition sequences that increase or enhance gene expression are often referred to as enhancers, whereas sequences that reduce or inhibit gene expression are called silencers.

environmental factors: In genetics, "environmental factors" has a specific meaning referring to all the factors that influence the expression of a trait other than genetic factors. Environmental factors can include diet, drug, or chemical exposures, or even prenatal development. [2]

epistasis: A situation in which two or more genes interact in a nonadditive way to produce a phenotype.

eukaryote: Organisms whose cells contain membrane-bound organelles including the nucleus, mitochondria, Golgi apparatus, and chloroplasts in plants. Organisms whose cells lack membrane-bound structures are referred to as prokaryotes.

gene deletion: A specific insertion–deletion mutation where the genetic sequence deleted or missing includes a functioning gene.

gene pool: All of the alleles for a given gene in a population make up the gene pool, with each allele having a specific allele frequency.

genetic background: The totality of all the genes within an individual. Given gene–gene interactions, the expression of any allele will be affected by the other alleles processed by an individual. The expression of an allele may change if the other alleles with which it interacts are different.

genetic marker: A region of DNA of known location in the genome that is polymorphic (has allelic differences). These alleles could be SNPs or indels in the DNA.

genome: The genome of an organism encompasses all the genetic material in the cell. In humans, this includes the 3 billion base pairs contained in the chromosomes in the nucleus and the approximately 16,000 base pairs of the mitochondrion.

genome-wide association studies (GWAS): Genetic studies that examine the entire genome for correlations among a very large number of known genetic markers and specific phenotypes.

genotype: The underlying genetic constitution of an individual, usually in relation to a specific trait. Genotyping refers to tests carried out to determine an individual's genotype.

HapMap Project: An international effort to identify and map the haplotype blocks of the human genome. Haplotype blocks are large segments of chromosomes whose sequence is nearly identical to that of many members of a population. The aim of the HapMap effort is to describe common patterns of human variation that arise due to linkage. In this way, large blocks of the genome (thousands of kilobases) can be characterized by a limited number of diagnostic SNPs (tag SNPs).

hemizygous: Cases in diploid cells where there is only one copy of a chromosome or chromosomal region. Males having only one X chromosome are said to be hemizygous for the sex chromosomes.

heritability: The proportion of total phenotypic variance due to genetic variation.

Hippocratic ethics: A system of medical ethics based on the so-called Oath of Hippocrates and asserting six fundamental principles for the practice of medicine: (1) respect for one's teachers and consideration toward one's colleagues, (2) the command "first, do no harm," (3) respect for the patient's life, (4) restriction of one's practice to one's field of expertise, (5) refraining from abuse of privilege, and (6) respect for the patient's privacy.

homo/heteropolymer: In cases where more than one polypeptide forms the protein, the polypeptides may be identical (homopolymer) or different (heteropolymer).

homozygous: A locus or individual is said to be homozygous if the two alleles present are identical. Heterozygous individuals carry different alleles at the locus of interest.

housekeeping genes: Genes whose transcription rate is relatively constant with differing cellular conditions. Often the protein products of housekeeping genes are needed for typical maintenance of the cell.

indel: Mutations defined by either the Insertion or Deletion of nucleotides in the genome. Indels can range in size from 1 nucleotide to millions of nucleotides. Includes copy number variants and gene deletions.

injustice: In medical ethics, the lack of fairness or justice that occurs when individuals who are equal in all medically relevant respects are treated differently on the basis of characteristics such as age, race, gender, or other non–medically relevant factors.

intergenic: Regions; that is, the space between neighboring genes on a chromosome.

justice: In medical ethics, the precept that all people, regardless of age, ethnicity, gender, national origin, race, religion, sexual orientation, social or economic status, or any other attribute or characteristic, should be treated equally and should receive the same quality of medical care.

linkage disequilibrium: Two genetic loci are said to be in linkage disequilibrium if they co-occur more frequently than would be expected given mendelian independent assortment. Two loci that are physically near one another on the chromosome are likely to be transmitted together; they are linked.

meiosis: Process during gamete formation in which the numbers of chromosomes per cell are divided in half. Prior to this reduction in genetic material, the DNA must be replicated.

monogenic: A trait whose expression is largely determined by a single gene. Traits of phenotypes affected by multiple genes are said to be polygenic.

mutation: A change in the DNA sequence of the genome. Mutations occurring in the germ line are potentially heritable. Changes in DNA sequence are of two basic types: single-nucleotide polymorphisms (SNPs) or insertion/deletions (indels) that can be from one to millions of nucleotides in size.

nitrogenous bases: For a DNA strand or polymer, the basic subunit is a nucleotide that consists of a 5-carbon sugar (deoxyribose) covalently bound to one of four possible nitrogenous bases: adenine (A), cytosine (C), guanine (G), or thymine (T).

nonmaleficence: An ethical principle holding that the health care provider should refrain from a course of action that would bring harm to the patient.

nucleoside: A deoxyribose sugar, with its attached nitrogeneous base, is called a *nucleoside*.

nucleotide: A nucleoside becomes a *nucleotide* with the addition of a phosphate group.

null allele: A variant allele that produces no gene product.

orthologous: Genes found in another organism that is homologous to the gene being studied in humans. Homologous genes share a common ancestry and are descended from the same ancestral gene. Members of the same gene family that have different functions in the cell and are not homologous are paralogs.

palindromic: A sequence on a DNA strand that has the same sequence on the forward strand as on the reverse strand.

pharmacogenetics: The study of the role of genetic variation in determining individual drug response. Generally, pharmacogenetics has been limited to the effects of one or a few genes.

pharmacogenomics: The study of the genome-wide role of human variation in drug response. Pharmacogenomics is a broad term that includes pharmaco-genetic effects. Pharmacogenomics also includes the application of genomic technologies in drug discovery, disposition, and function.

phenocopies: Traits that are caused by environmental factors and are indistinct from the trait being examined for genetic causes.

phenotype: The observable outcome of the interaction of an individual's genes and environmental factors.

polygenic: Traits whose outcome or phenotype are due to the action of two or more genes.

polymorphic: A gene or locus is polymorphic if there are differences among individuals in its DNA sequence or length. Generally, the specific difference must have a frequency of 5% in the population to be considered polymorphic.

promoter region: A DNA sequence immediately adjacent to and largely upstream of a gene. Sequences located here facilitate the binding of RNA polymerase and proteins (transcription factors) necessary for transcription.

pseudogenes: Nonfunctional genes in the genome that are no longer expressed in any member of the species.

rare allele effect: A theory to explain how recessive deleterious, even lethal alleles can be maintained in a population in spite of strong selection to remove them. As selection acts to remove a deleterious allele from a population, the frequency of the allele becomes even rarer. Because selection can act only on the homozygous individuals for a recessive allele, the opportunity to remove the deleterious allele is exceedingly rare. At the same time the deleterious allele is hidden from selection in the heterozygotes.

recessive: A property of one of two alleles. An allele is said to be recessive when its phenotype is masked or unseen when in combination with another allele. The other allele is said to be dominant. Recessive alleles need not be rare in a population nor deleterious to the individual.

reverse transcriptase: An enzyme capable of synthesizing a DNA strand using an RNA template. Retroviruses are the only known entities that have functional reverse transcriptase. Reverse transcriptase is a fundamental tool in molecular biology research for the conversion of mRNA into its complementary DNA (cDNA). A step necessary for DNA sequencing, amplification in PCR, or cloning.

RNA interference: A cellular mechanism of gene silencing mediated by short double-stranded RNA (dsRNA) molecules that are taken up by the RNA-induced silencing complex with the cytoplasm of the cell and together inhibit the translation of specific mRNAs based upon the sequence of the short dsRNA.

semi-dominance: Pattern of inheritance where the heterozygote has an intermediate phenotype. For example, cases of semi-dominance may arise when one of the alleles of a gene results in loss of expression. Heterozygotes would potentially produce half the gene product, which may be detectable as reduced function.

stigmatization: The negative characterization or categorization of patients on the basis of their genetic profile or other characteristics, such as age, race, or gender.

stratification: In a clinical trial, the process of classifying individuals on the basis of distinguishing characteristics, such as age, sex, race, lifestyle, socioeconomic status, and genetic profile.

transcription: The process of synthesizing an RNA strand based upon a DNA template. This is the first step in the expression of a gene. The enzyme responsible for the synthesis of the RNA is called RNA polymerase.

transcription factors: Special proteins that, upon binding, regulate gene transcription or expression. Transcription factors are coded for by genes that are not adjacent to the genes they regulate and therefore referred to as trans-elements or factors. The sequences of DNA that bind transcription factors generally near the genes being regulated are referred to as cis-elements or cis-acting.

transcriptome: The set of all RNAs in a cell. These include mRNAs, tRNAs, and rRNAs, though generally mRNAs are the species of interest because mRNAs give rise to proteins.

transgene: A gene or genetic material that has been transferred from one species into another.

vector: Vehicles used to move DNA fragments from one entity to another, for example, from one cell to a target cell. Depending upon the need, vectors can be plasmids, viruses, or artificial chromosomes.

Watson-Crick rule: In a DNA molecule, adenosine specifically binds to thymine via two hydrogen bonds, and cytosine binds to guanine via three hydrogen bonds.

whole-genome duplication: Rare mutational event in which the chromosome number is doubled or fails to be reduced during cell division. Much more common are subgenomic duplication events where smaller chromosomal regions are duplicated. Often these smaller regions contain one or more genes.

Index

Page numbers followed by *f* or *t* indicate material in figures or tables, respectively.

A

E

efficacy, drug, 2, 2*f*, 4
electronic medical records, 87
endonucleases, 54–55, 56*f*
enhancers, 18
environmental factors, 22, 30, 71–72
enzymes
    drug-metabolizing, 6, 41, 48–50. *See also* cytochrome P450 gene family
    restriction, 54–55, 56*f*
    structure of, heritable differences in, 5
epistasis, 25, 26*f*, 31
Escherichia coli, genome of, 37*t*
Ethical Principles and Guidelines for the Protection of Human Subjects of Research, 82–85
ethics, 81–96
    basic principles of, 82
    Belmont Report on, 82–85
    challenges for pharmacists, 81–82
    Hippocratic, 83
    implications for patients, 85–87
    implications for pharmacists and health care professionals, 87–89
    individuals and groups involved in issues, 82, 82*t*
    in personalized medicine, 84–85
ethnicity, 92, 93*f*
eukaryotes
    chromosomes of, 38
    definition of, 37
    genomes of, 36–37
European Molecular Biology Laboratory, 65
evolution
    of gene families, 41–43
    of genomes, 36
exons, 16, 17*f*, 43
exonuclease, 16
extensive metabolizers, 6

F

factor VIII gene, 45*t*
family studies, 28–29, 28*f*
fast acetylators, 5
fava beans/favism, 5

Principles of the Human Genome and Pharmacogenomics

nitrogenous bases, 12–15, 13*f*
    pairing of, 14, 14*f,* 16
    Watson-Crick rule for, 13, 14, 16
noncoding DNA, 37
noncoding RNA, 16
nonmaleficence, 84
nonmendelian inheritance, 26–28
nonpolar amino acids, 15, 15*t*
nuclear genome, 35, 37–38
nucleoside, 12, 13
nucleotides, 12–13, 13*f*
null allele, 25

O

off-label prescribing, 89
O-glycosylation, 21
olfactory receptor (OR) genes, 41, 41*f*
Online Mendelian Inheritance in Man (OMIM), 65–66
OR genes. *See* olfactory receptor genes
orphan drugs, 90–91
orthologous genes (orthologs), 50
oxidative phosphorylation, 37

P

palindromic sequences, 55, 59
patients
    autonomy of, 86
    confidentiality for/privacy of, 86–87, 92
    ethical implications for, 85–87
    ethical principles on, 82–85
    individual genomes of, 7, 59
    personalized medicine for, 7–8, 59, 70, 84–85
    respect for, 82
    stigmatization of, 85, 86, 92
PCR. *See* polymerase chain reaction
pedigree
    for mendelian inheritance, 23–24, 24*f*
    for nonmendelian inheritance, 28*f*
personalized medicine, 7–8, 70
    ethics in, 84–85
    individual genomes in, 7, 59

T